Lecture Notes in Mathematics

Edited by A. Dold and B. Ec

911

Ole G. Jørsboe
Leif Mejlbro

The Carleson-Hunt Theorem on Fourier Series

Springer-Verlag
Berlin Heidelberg New York 1982

Authors

Ole Groth Jørsboe
Leif Mejlbro
Department of Mathematics, Technical University of Denmark
DK-2800 Lyngby, Denmark

AMS Subject Classifications (1980): 43A50

ISBN 3-540-11198-0 Springer-Verlag Berlin Heidelberg New York
ISBN 0-387-11198-0 Springer-Verlag New York Heidelberg Berlin

Printing and binding: Beltz Offsetdruck, Hemsbach/Bergstr.
2141/3140-543210

CONTENTS.

PREFACE.

The main purpose of this monograph is to give a selfcontained proof of the famous Carleson-Hunt theorem (cf. [1] and [4] and [7]), i.e. if $f \in L^p(]-\pi,\pi])$, $p \in]1,+\infty]$, then $S_n(x;f) \to f(x)$ for almost every $x \in]-\pi,\pi]$ as $n \to +\infty$, where $S_n(x;f)$ denotes the n-th partial sum of the Fourier series for f . Except for Lebesgue's differentiation theorem, which we shall take for granted (cf. e.g. [3]), we shall only need elementary Measure Theory. This means that the exposition has become fairly long, as we shall need some classical results concerning Marcinkiewicz' interpolation theorems and the Hilbert transform.

The text is composed of four chapters. In chapter I we shall prove some interpolation theorems and prove Carleson-Hunt's theorem under the assumption that some operator M is of type p for $p \in]1,+\infty[$, so the rest of the book is concerned with the proof of this assumption. Chapter II is dealing with the Hilbert transform, and we prove some exponential estimates needed in the following. Chapter III is more technical as we introduce the dyadic intervals and the generalized Fourier coefficients. In order to avoid some possible misunderstandings we have changed the usual notation. Except for the dyadic intervals (which in some sense are quite reasonable) the results are still fairly general. Finally, in chapter IV, we shall concentrate on the given situation. This chapter is very difficult, which was to be expected, as the Carleson-Hunt theorem is far from obvious. We shall prove the existence of an exceptional set, in which we cannot be sure that the Fourier series converges pointwise towards $f(x)$, and we shall at last prove that this exceptional set is in fact a null-set and thus prove the theorem.

CHAPTER I.

This chapter is composed of four sections. In § 1 we introduce the concept of (weak and strong) type of an operator, and we prove an interpolation theorem, which is a special case of a theorem due to Marcinkiewicz (cf. [9] for the general formulation). In § 2 we introduce the Hardy-Littlewood maximal operator Θ and prove that Θ is of type p for all $p \in]1,+\infty[$. In § 3 another classical interpolation theorem is proved, namely the Stein-Weiss theorem, and finally, in § 4, we prove the Carleson-Hunt theorem under the assumption that some operator M defined below is of type p for all $p \in]1,+\infty[$.

For technical reasons we shall always consider *real-valued* functions defined on a finite interval, although their Fourier expansions will be written by means of the complex exponential functions. This assumption will save us for a lot of trouble in the estimates in the following chapters, and we do not loose any generality, since for a complex-valued function f we may consider the two real-valued functions $\operatorname{Re} f$ and $\operatorname{Im} f$ instead. We shall further assume that f is integrable , $f \in L^1(I)$, where I is the finite interval mentioned above. (We may note that most of the results in this chapter also hold for $f \in L^1(R)$, but their proofs may be different and more complicated, so we have avoided to prove the theorems in their full generality.)

§ 1. Interpolation theorems.

Let f be a real-valued function defined on an interval $[-A,A]$ and suppose that $f \in L^1([-A,A])$.

The Lebesgue-measure on $\mathbb{R}$ is denoted by m . We introduce the sets E_y depending on the function f under consideration by

$$(1.1) \qquad E_y = \{x \in [-A,A] \mid |f(x)| > y\} , \quad y \in \mathbb{R}_+ .$$

Definition 1.1. *By the* *distribution function* $\lambda_f(y)$, $y \in \mathbb{R}_+$, *we shall understand the function* $\lambda_f : \mathbb{R}_+ \to [0,2A]$ *defined by*

$$(1.2) \qquad \lambda_f(y) = m(E_y) = m(\{x \in [-A,A] \mid |f(x)| > y\}) .$$

Clearly, we have $0 \leqq \lambda_f(y) \leqq 2A$ for all $y \in \mathbb{R}_+$, $\lambda_f(y) \to 0$ as $y \to +\infty$, and $\lambda_f(y)$ is a decreasing function. Furthermore, λ_f is continuous from the right, as

$$\bigcup_{n=1}^{+\infty} E_{y+\frac{1}{n}} = E_y \quad \text{for all} \quad y \in \mathbb{R}_+ .$$

From this fact we conclude that λ_f has at most countably many discontinuity points, and especially, λ_f is a measurable function.

Let T be an operator from $L^1([-A,A])$ into $\mathfrak{M}([-A,A])$, the set of all measurable functions on $[-A,A]$. The operator T will not necessarily be defined on all of $L^1([-A,A])$, but will at least be defined on all *simple functions* , i.e. finite linear combinations of indicator functions of measurable subsets of $[-A,A]$, and on all continuous functions, and so especially the domain of T , $\mathcal{D}(T)$, is dense in $L^1([-A,A])$.

In the following T will either be a *linear operator* [i.e. $\mathcal{D}(T)$ is a linear set and $T(\alpha f + \beta g) = \alpha Tf + \beta Tg$ for all $f, g \in \mathcal{D}(T)$ and all $\alpha, \beta \in \mathbb{R}(\mathbb{C})$] , or T will be a *sublinear operator* [i.e. $\mathcal{D}(T)$ is a linear set, $|T(\alpha f)| = |\alpha|\,|Tf|$ for all $f \in \mathcal{D}(T)$ and all $\alpha \in \mathbb{R}(\mathbb{C})$, and $|T(f+g)| \leqq |Tf| + |Tg|$ for all $f, g \in \mathcal{D}(T)$] .

Definition 1.2. *The operator* T *is said to be of* (*strong*) *type* p (*where* $p \in [1,+\infty]$) *if there exists a constant* $A_p \in R_+$, *such that*

$$(1.3) \qquad \|Tf\|_p \leq A_p \|f\|_p \quad \text{for } f \in \mathfrak{D}(T) .$$

Remark 1.3. Usually one considers operators T of (strong) type (p,q), where $p,q \in [1,+\infty]$. The operator T is said to be of (strong) type (p,q), if there exists a constant $A_{p,q} \in R_+$, such that

$$\|Tf\|_q \leq A_{p,q} \|f\|_p \quad \text{for } f \in \mathfrak{D}(T) .$$

(Cf. e.g. [9] and [7]). For our purpose it is enough to consider the case, where $p = q$. Analogous remarks are applicable to the definitions in the following.

We note that if an operator T is of type p with $p \in [1,+\infty[$, then T can be extended to all of $L^p([-A,A])$ by continuity, since $\mathfrak{D}(T)$ is dense in $L^p([-A,A])$, and T is thus a bounded operator defined on all of $L^p([-A,A])$.

Definition 1.4. *The operator* T *is said to be of* weak type p (*where* $p \in [1,+\infty[$), *if there exists a constant* $A_p \in R_+$, *such that*

$$(1.4) \qquad \lambda_{Tf}(y) \leq \left(\frac{A_p}{y}\right)^p \|f\|_p^p$$

for all $f \in \mathfrak{D}(T)$ *and for all* $y \in R_+$.

Note that for each $y \in R_+$,

$$\|Tf\|_p^p = \int |Tf(x)|^p dx \geq y^p \lambda_{Tf}(y) .$$

Hence we conclude that if T is of type p, $p \in [1,+\infty[$, then

$$(1.5) \qquad \lambda_{Tf}(y) \leq \frac{1}{y^p} \|Tf\|_p^p \leq \left(\frac{A_p}{y}\right)^p \|f\|_p^p ,$$

i.e. T is also of weak type p.

We shall later on give an example showing that the converse is not necessarily true.

We shall also need the following concepts, closely related to what we already have introduced.

Definition 1.5. *The operator* T *is said to be of* restricted type p *(where* $p \in [1,+\infty[$ *) if there exists a constant* $A_p \in \mathbb{R}_+$ *, such that*

$$\|T\chi_E\|_p \leqq A_p\|\chi_E\|_p = A_p[m(E)]^{1/p} \tag{1.6}$$

for all measurable sets $E \subseteq [-A,A]$.

Definition 1.6. *The operator* T *is said to be of* restricted weak type p *(where* $p \in [1,+\infty[$ *) if there exists a constant* $A_p \in \mathbb{R}_+$ *, such that*

$$\lambda_{T\chi_E}(y) \leqq \left(\frac{A_p}{y}\right)^p \|\chi_E\|_p^p = \left(\frac{A_p}{y}\right)^p m(E) \tag{1.7}$$

for all measurable sets $E \subseteq [-A,A]$.

Clearly, if T is of type p then T is also of restricted type p , and if T is of weak type p then T is also of restricted weak type p . Furthermore, if T is of restricted type p then T is also of restricted weak type p . (This is proved in quite the same way as the assertion following definition 1.4.)

Before we start proving the interpolation theorems we need a result relating the distribution function λ_f to the p-norm $\|f\|_p$ of f . In fact, we have

Lemma 1.7. *If* $f \in L^1([-A,A])$ *then for each* $p \in [1,+\infty[$

$$\|f\|_p^p = \int |f(x)|^p\,dx = \int_0^{+\infty} p\,y^{p-1}\lambda_f(y)dy\ , \tag{1.8}$$

and especially,

$$\|f\|_1 = \int_0^{+\infty} \lambda_f(y)dy\ . \tag{1.9}$$

Proof. Using Fubini's theorem we have

$$\int |f(x)|^p dx = \int \left(\int_0^{|f(x)|} p\, y^{p-1} dy \right) dx$$

$$= \int_0^{+\infty} p\, y^{p-1} m(\{x \in [-A,A] \mid |f(x)| > y\}) dy = \int_0^{+\infty} p\, y^{p-1} \lambda_f(y) dy \; .$$

Note that if $f \notin L^p$ then both sides of (1.8) are $+\infty$. □ .

We now prove the first interpolation result.

Lemma 1.8. *Assume that* T *is of restricted weak type* p_o *and* p_1 *, where* $1 \leq p_o < p_1 < +\infty$ *. Then* T *is of restricted type* p *for all* $p \in]p_o,p_1[$ *.*

Proof. Let $\lambda(y)$ denote the distribution function of $T\chi_E$. From the assumptions we know that there exists constants A_o and A_1 , such that

$$\lambda(y) \leq \left(\frac{A_o}{y}\right)^{p_o} m(E) \; , \qquad \lambda(y) \leq \left(\frac{A_1}{y}\right)^{p_1} m(E)$$

for all measurable sets E and all $y \in R_+$. Using lemma 1.7 we get

$$\|T\chi_E\|_p^p = p \int_0^1 y^{p-1} \lambda(y) dy + p \int_1^{+\infty} y^{p-1} \lambda(y) dy$$

$$\leq p \cdot m(E) \left\{ \int_0^1 y^{p-p_o-1} A_o^{p_o} dy + \int_1^{+\infty} y^{p-p_1-1} A_1^{p_1} dy \right\}$$

$$= p \cdot m(E) \left\{ A_o^{p_o} \cdot \frac{1}{p-p_o} + A_1^{p_1} \cdot \frac{1}{p_1-p} \right\}$$

and thus

$$\|T\chi_E\|_p \leq A_p [m(E)]^{1/p}$$

with

$$A_p = p^{1/p} \cdot \left\{ A_o^{p_o} \cdot \frac{1}{p-p_o} + A_1^{p_1} \cdot \frac{1}{p_1-p} \right\}^{1/p} \quad \text{for } p \in]p_o,p_1[\; . \qquad \square$$

We note that A_p is bounded as long as $p \in]p_o, p_1[$ is bounded away from p_o and p_1.

A closely related result is the following which is a special case of a theorem due to Marcinkiewicz (cf. [9]).

Theorem 1.9. *Let T be a sublinear operator of weak type p_o and p_1, where $1 \leqq p_o < p_1 < +\infty$. Then T is of type p for all $p \in]p_o, p_1[$. More precisely we have that if*

$$\lambda_{Tf}(y) = m(\{x \mid |Tf(x)| > y\}) \leqq \left(\frac{A_o}{y}\right)^{p_o} \|f\|_{p_o}^{p_o} ,$$

$$\lambda_{Tf}(y) = m(\{x \mid |Tf(x)| > y\}) \leqq \left(\frac{A_1}{y}\right)^{p_1} \|f\|_{p_1}^{p_1} ,$$

then

$$\|Tf\|_p^p \leqq K_p \|f\|_p^p ,$$

where

$$K_p = p \cdot 2^p \left(\frac{1}{p-p_o} + \frac{1}{p_1-p}\right) A_o^{p_o \cdot \frac{p_1-p}{p_1-p_o}} \cdot A_1^{p_1 \cdot \frac{p-p_o}{p_1-p_o}} .$$

Proof. We choose $A = A_o^{\frac{p_o}{p_1-p_o}} A_1^{\frac{-p_1}{p_1-p_o}}$ and note for later use that

$$A_o^{p_o} A^{p_o-p} = A_1^{p_1} A^{p_1-p} = A_o^{p_o \frac{p_1-p}{p_1-p_o}} \cdot A_1^{p_1 \frac{p-p_o}{p_1-p_o}} . \tag{1.10}$$

Using the constant A defined above we introduce the functions f^y and f_y for fixed $y \in R_+$ by

$$f^y(x) = \begin{cases} f(x) & \text{if } |f(x)| \leqq Ay \\ 0 & \text{otherwise} \end{cases} , \qquad f_y(x) = \begin{cases} 0 & \text{if } |f(x)| \leqq Ay \\ f(x) & \text{otherwise} \end{cases} .$$

Clearly, we have $f(x) = f_y(x) + f^y(x)$, and the sublinearity of T gives us that

$$\lambda_{Tf}(2y) \leqq \lambda_{Tf_y}(y) + \lambda_{Tf^y}(y) ,$$

which by assumption is smaller than

$$A_o^{p_o} y^{-p_o} \int |f_y(x)|^{p_o} dx + A_1^{p_1} y^{-p_1} \int |f^y(x)|^{p_1} dx \quad .$$

Using Fubini's theorem to interchange the order of integration, and the constant A introduced above, we get

$$\|Tf\|_p^p = \int_0^{+\infty} p(2y)^{p-1} \lambda_{Tf}(2y)d(2y) = p \cdot 2^p \int_0^{+\infty} y^{p-1} \lambda_{Tf}(2y)dy$$

$$\leqq p \cdot 2^p \int_0^{+\infty} A_o^{p_o} y^{-p_o} y^{p-1} \left(\int_{\{|f(x)|>Ay\}} |f(x)|^{p_o} dx \right) dy$$

$$+ \int_0^{+\infty} A_1^{p_1} y^{-p_1} y^{p-1} \left(\int_{\{|f(x)|\leqq Ay\}} |f(x)|^{p_1} dx \right) dy \Big\}$$

$$= p \cdot 2^p \left\{ A_o^{p_o} \int |f(x)|^{p_o} \left(\int_{y=o}^{\frac{|f(x)|}{A}} y^{p-p_o-1} dy \right) dx \right.$$

$$\left. + A_1^{p_1} \int |f(x)|^{p_1} \left(\int_{\frac{|f(x)|}{A}}^{+\infty} y^{p-p_1-1} dy \right) dx \right\}$$

$$= p \cdot 2^p \left\{ A_o^{p_o} A^{p_o-p} \cdot \frac{1}{p-p_o} \int |f(x)|^p dx + A_1^{p_1} A^{p_1-p} \cdot \frac{1}{p_1-p} \int |f(x)|^p dx \right\}$$

$$= p \cdot 2^p \cdot A_o^{p_o \cdot \frac{p_1-p}{p_1-p_o}} \cdot A_1^{p_1 \cdot \frac{p-p_o}{p_1-p_o}} \cdot \left\{ \frac{1}{p-p_o} + \frac{1}{p_1-p} \right\} \|f\|_p^p ,$$

where we have used (1.10). □

In a similar way we get the following result.

Theorem 1.10. *Let* T *be a sublinear operator of weak type* p_o *and of strong type* $+\infty$ *, where* $p_o \in [1, +\infty[$ *. Then* T *is of type* p *for all* $p \in]p_o, +\infty[$ *. More precisely we have that if*

$$\lambda_{Tf}(y) = m(\{x \mid |Tf| > y\} \leqq \left(\frac{A_o}{y}\right)^{p_o} \|f\|_{p_o}^{p_o} \quad \text{and} \quad \|Tf\|_\infty \leqq A_\infty \|f\|_\infty ,$$

then

$$\|Tf\|_p^p \leqq p \cdot 2^p \cdot A_o^{p_o} A_\infty^{p-p_o} \cdot \frac{1}{p-p_o} \|f\|_p^p .$$

Proof. We use the same notation as in the proof of theorem 1.9. We choose the constant A as $\frac{1}{A_\infty}$. Then $\|f^y\|_\infty \leqq \frac{1}{A_\infty} y$, $\|Tf^y\|_\infty \leqq y$, and consequently $\lambda_{Tf^y}(y) = 0$. This gives the estimate

$$\lambda_{Tf}(2y) \leqq \lambda_{Tf_y}(y) \leqq A_o^{p_o} y^{-p_o} \int |f_y(x)|^{p_o} dx ,$$

and (exactly as in the proof of theorem 1.9)

$$\|Tf\|_p^p \leqq p \cdot 2^p \cdot A_o^{p_o} \cdot \left(\frac{1}{A_\infty}\right)^{p_o-p} \cdot \frac{1}{p-p_o} \|f\|_p^p ,$$

proving the theorem. □

§ 2. The Hardy-Littlewood maximal operator.

In this section we shall consider the Hardy-Littlewood maximal operator and derive estimates for this operator using theorem 1.10.

Let $f \in L^1(R)$. We define the maximal operator Θ by

$$\text{(2.1)} \qquad \Theta f(x) = \sup_{t \in R_+} \frac{1}{2t} \int_{x-t}^{x+t} |f(y)| dy , \quad x \in R .$$

Clearly, the function Θf is measurable for each $f \in L^1(R)$, because Θf is lower semicontinuous as a supremum of continuous functions, and the operator Θ is a sublinear operator, defined on $L^1(R)$.

Theorem 2.1. *The operator* Θ *is of type* $+\infty$ *and of weak type* 1 *. More precisely,* Θ *satisfies the following estimates*

(2.2) $$\|\Theta f\|_\infty \leq \|f\|_\infty ,$$

(2.3) $$\lambda_{\Theta f}(y) = m(\{x \mid \Theta f(x) > y\}) \leq \frac{4}{y}\|f\|_1 \quad \text{for all} \quad y \in R_+ .$$

From theorem 2.1 and theorem 1.10 we infer at once that the operator Θ is of type p for all $p \in]1,+\infty[$. In the applications later on we only need functions of compact support, so let us assume for the proof that f has compact support. It is obvious that (2.2) is satisfied. In the proof of (2.3) we may as well assume that f is non-negative. Let $y \in R_+$. To each $x \in \{t \mid \Theta f(t) > y\}$ there exists an interval I_x (with center at x) such that

$$\int_{I_x} f(t)dt > y \cdot m(I_x) .$$

As f has compact support, the set $\{\Theta f > y\}$ is bounded, and we may suppose that all the intervals I_x are contained in an interval $[-B,B]$.

We claim the existence of a sequence (I_n) (extracted from the class of intervals I_x above) of pairwise disjoint intervals, such that

$$m\left(\bigcup_{n=1}^{+\infty} I_n\right) \geq \frac{1}{4} m\left(\bigcup_x I_x\right) .$$

Assume for the moment that this has been proved. Then the proof is finished as follows:

$$m(\{x \mid \Theta f(x) > y\}) \leq m\left(\bigcup_x I_x\right) \leq 4m\left(\bigcup_{n=1}^{+\infty} I_n\right) = 4\sum_{n=1}^{+\infty} m(I_n)$$

$$\leq \frac{4}{y}\sum_{n=1}^{+\infty}\int_{I_n} f(y)dy = \frac{4}{y}\int_{\bigcup_n I_n} f(y)dy \leq \frac{4}{y}\|f\|_1 .$$

Let us prove the assertion above (this is a theorem of Besicovitch-type; results in that direction can be found in e.g. [3])

Let S_1 be the class of all intervals I_x considered above. Let $a_1 = \sup\{m(I_x) \mid I_x \in S_1\}$ and choose an interval I_1 from S_1 with $m(I_1) > \frac{3}{4}a_1$. Let S_2 be the class of intervals I_x that do not intersect I_1 , let $a_2 = \sup\{m(I_x) \mid I_x \in S_2\}$ and choose I_2 from S_2 with $m(I_2) > \frac{3}{4}a_2$. In this way we continue. If the process stops after a finite number of steps, the result is obvious, so we may assume that we obtain a sequence (a_k) of real numbers $a_k \to 0$ and a corresponding sequence of disjoint intervals (I_k) with $m(I_k) \to 0$ for $k \to +\infty$.

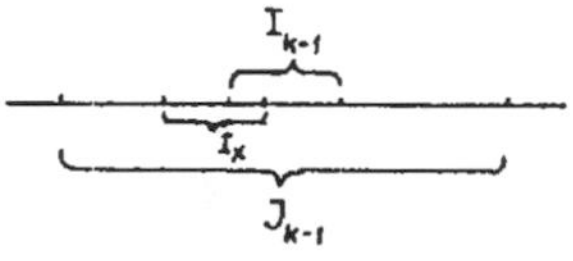

Let us consider an interval I_x and let k denote the first index for which $I_x \notin S_k$; in that case $I_x \cap I_{k-1} \neq \emptyset$ and $m(I_{k-1}) \geqq \frac{3}{4} m(I_x)$. Then we have $I_x \subset J_{k-1}$ where J_{k-1} and I_{k-1} have the same center and where $m(J_{k-1}) = 4m(I_{k-1})$. Using this fact we get $m\left(\bigcup_x I_x\right) \leqq 4m\left(\bigcup_{n=1}^{+\infty} I_n\right)$, and the theorem is proved. It should be noted that the constant 4 appearing in (2.3) is not the best possible one. $\square$.

Note that Θ is *not* of type 1 (if we e.g. let $f(x) = \chi_{]0,1[}(x)$, then Θ is not even integrable!)

<u>Corollary 2.2.</u> *The operator* Θ *is of type* p *for all* $p \in]1,+\infty[$. *More precisely we have*

$$(2.4) \qquad \|\Theta f\|_p^p \leqq 4p \cdot 2^p \cdot \frac{1}{p-1} \|f\|_p^p \quad \text{for all } p \in]1,+\infty[.$$

As mentioned above this follows from theorem 2.1 using theorem 1.10 with $p_0 = 1$, $A_0 = 4$ and $A_\infty = 1$. $\square$.

If $g(x)$ is a real-valued function, we define another function $g^+(x)$ by $g^+(x) = \max\{g(x),0\}$. A well-known example is $\log^+$.

<u>Theorem 2.3.</u> *The operator* Θ *satisfies*

$$(2.5) \qquad \int (\Theta f(x) - 2)^+ dx \leqq 8 \int |f(x)| \log^+ |f(x)| dx .$$

Proof. We may of course assume that $f \geqq 0$ and that $f \log^+ f$ is integrable. We define functions f^y and f_y by

$$f^y(x) = \begin{cases} f(x) & \text{if } f(x) \leqq y \\ 0 & \text{otherwise} \end{cases}, \qquad f_y(x) = \begin{cases} 0 & \text{if } f(x) \leqq y \\ f(x) & \text{otherwise} \end{cases}$$

Then clearly $\Theta f^y \leqq y$ and so $m(\{x \mid \Theta f(x) > 2y\}) \leqq m(\{x \mid \Theta f_y(x) > y\})$. Using (2.3) on the function f_y we get

$$m(\{x \mid \Theta f(x) > 2y\}) \leqq \frac{4}{y} \int f_y(x)dx \,. \tag{2.6}$$

An integration of (2.6) from 1 to $+\infty$ gives

$$\int_1^{+\infty} m(\{x \mid \Theta f(x) > 2y\})dy \leqq \int_1^{+\infty} \frac{4}{y} \left(\int f_y(x)dx\right) dy$$

$$= 4 \int_{\{|f(x)|>1\}} \left(\int_{y=1}^{|f(x)|} \frac{1}{y} dy\right) dx = 4 \int f(x) \log^+ f(x)dx \,.$$

The left hand side equals $\frac{1}{2}\int_2^{+\infty} m(\{x \mid \Theta f(x) > t\})dt = \frac{1}{2}\int (\Theta f(x) - 2)^+ dx$, and thus the result follows. □

Later on we shall also need the following two simple lemmata:

Lemma 2.4. *Let $f \in L^1(\mathsf{R})$. For any fixed $x \in \mathsf{R}$ we define a function F_x on R by $F_x(t) = \int_0^t f(x+y)dy$. Then*

$$|F_x(t)| \leqq 2|t|\Theta f(x) \qquad \textit{for all } x \in \mathsf{R} \textit{, all } t \in \mathsf{R}$$

Proof. The lemma follows immediately from

$$|F_x(t)| = \left|\int_x^{x+t} f(y)dy\right| \leqq \left|\int_x^{x+t} |f(y)|dy\right| \leqq \int_{x-|t|}^{x+|t|} |f(y)|dy \leqq 2|t| \cdot \Theta f(x) \,. \quad \square$$

Lemma 2.5. *If $f \in L^1(\mathsf{R})$, then*

$$|f(x)| \leqq \Theta f(x) \qquad \textit{for almost every } x \in \mathsf{R} \,.$$

Proof. The lemma follows from the following two facts

$$\Theta f(x) \geqq \frac{1}{2t}\int_{x-t}^{x+t} |f(y)|dy \quad \text{for all} \quad t \in R_+$$

and

$$\frac{1}{2t}\int_{x-t}^{x+t} |f(y)|dy \to |f(x)| \quad \text{for} \quad t \to 0+ \quad \text{for almost every} \quad x \in R ,$$

where we have used Lebesgue's differentiation theorem for L^1-functions (cf. e.g. [3]). □

From lemma 2.5 and theorem 2.1 follows that if $f \in L^1(R) \cap L^\infty(R)$ we even get $\|f\|_\infty = \|\Theta f\|_\infty$.

Finally we shall prove an exponential estimate for Θf , when $f \in L^\infty$ and $f(x)$ is zero outside a compact set.

Theorem 2.6. *Let $c \in R_+$ be any positive constant and let f be any essentially bounded function, the support of which is contained in an interval I of length A . Then for $y \in R_+$*

$$(2.7) \qquad \lambda_{\Theta f}(y) = m(\{x \mid \Theta f(x) > y\}) \leqq 2A \cdot \exp c \cdot \frac{\|f\|_\infty}{y} \cdot \exp\left(-c \cdot \frac{y}{\|f\|_\infty}\right) .$$

Proof. It follows from theorem 2.1 that $\|\Theta f\|_\infty \leqq \|f\|_\infty$ (in this case we even have $\|\Theta f\|_\infty = \|f\|_\infty$) , so if $y \geqq \|f\|_\infty$ the left hand side of (2.7) is zero, and the estimate is trivial.

Let $y \in]0,\|f\|_\infty[$. If $x \in \{x \mid \Theta f(x) > y\}$ and $\operatorname{dist}(x,I) = d > 0$, then an application of (2.1) gives

$$y < \Theta f(x) \leqq \frac{1}{2d}\int_I |f(\tau)|d\tau \leqq \frac{A}{2d}\|f\|_\infty ,$$

from which we conclude that

$$m(\{x \mid \Theta f(x) > y\}) \leqq A + 2 \cdot \frac{A}{2}\frac{\|f\|_\infty}{y} = A \cdot \left(1 + \frac{\|f\|_\infty}{y}\right) .$$

Let $t = \frac{y}{\|f\|_\infty} \in]0,1[$. Then it is enough to prove that

$$1+\frac{1}{t} \leqq 2\exp c \cdot \frac{1}{t} \cdot \exp(-ct) \;, \quad t \in]0,1[\;,$$

which is equivalent to the trivial estimate

$$(1+t)\exp(ct) \leqq 2 \cdot \exp c \quad \text{for} \quad t \in]0,1[\;,$$

and the theorem is proved. □

§ 3. The Stein-Weiss theorem.

We shall now introduce an operator T^* associated with the operator T. In the following we assume that all functions considered are defined on a fixed interval $[-A,A]$. Let T be a *linear* operator of restricted type p, where $p \in]1,+\infty[$, i.e. there exists a constant $A_p \in R_+$ such that for every measurable set $E \subseteq [-A,A]$,

(3.1) $$\|T\chi_E\|_p \leqq A_p \|\chi_E\|_p = A_p [m(E)]^{1/p} .$$

Let q be the number conjugate to p, i.e. $\frac{1}{p}+\frac{1}{q}=1$, and let $f \in L^q([-A,A])$. We define a set function γ on the class of Borel sets E contained in $[-A,A]$ by

(3.2) $$\gamma(E) = \int (T\chi_E) f \, dx$$

Hölder's inequality implies that γ is well-defined, and it is easy to see that γ is a countably additive set function which is absolutely continuous with respect to Lebesgue measure. Using Radon-Nikodym's theorem we then get an (up to nullsets) uniquely determined function h such that

(3.3) $$\gamma(E) = \int_E h(x)dx = \int \chi_E \cdot h \, dx = \int T\chi_E \cdot f \, dx \;,$$

for all Borel sets $E \subseteq [-A,A]$.

Because of this relation we *define* an operator T^* on $L^q([-A,A]$ by

(3.4) $$T^*f = h \;.$$

Clearly, T^* is linear, and *formally* T^* behaves as the adjoint operator of T. We have e.g. that if g is a simple function, then

(3.5) $$\int_{-A}^{A} Tg \cdot f \, dx = \int_{-A}^{A} g \cdot T^*f \, dx \;.$$

Lemma 3.1. *Assume that the linear operator* T *is of restricted type* p, *where* $p \in]1,+\infty[$. *Then* T^* *is of weak type* q, *where* $\frac{1}{p}+\frac{1}{q}=1$.

Proof. By assumption there exists a constant $A_p \in \mathbb{R}_+$, such that

$$\|T\chi_E\|_p \leqq A_p\|\chi_E\|_p = A_p[m(E)]^{1/p} . \tag{3.6}$$

We shall prove the existence of a constant $B_q \in \mathbb{R}_+$, such that

$$\lambda_{T^*f}(y) \leqq \left(\frac{B_q}{y}\right)^q \|f\|_q^q \quad \text{for all } f \text{ and all } y \in \mathbb{R}_+ . \tag{3.7}$$

Let $f \in L^q$ and let $h = T^*f$ and $\lambda(y) = \lambda_h(y) = m(E_y)$, where as usual, $E_y = \{x \mid |h(x)| > y\}$. We put $E_y = E_y^+ \cup E_y^-$, where

$$E_y^+ = \{x \mid h(x) > y\} , \qquad E_y^- = \{x \mid h(x) < -y\} .$$

Finally, we define

$$\lambda^+(y) = m(E_y^+) , \qquad \lambda^-(y) = m(E_y^-) .$$

Clearly, $\lambda(y) = \lambda^+(y) + \lambda^-(y)$, and for $\lambda^+(y)$ we get the following estimate

$$y\,\lambda^+(y) = y\int \chi_{E_y^+}dx = \int \chi_{E_y^+}\cdot y\,dx \leqq \int \chi_{E_y^+}\cdot h\,dx = \int \chi_{E_y^+}\cdot T^*f\,dx$$

$$= \int T\chi_{E_y^+}\cdot f\,dx \leqq \|T\chi_{E_y^+}\|_p\,\|f\|_q \leqq A_p\|\chi_{E_y^+}\|_p\,\|f\|_q$$

$$= A_p[m(E_y^+)]^{1/p}\,\|f\|_q = A_q[\lambda^+(y)]^{1-(1/q)}\,\|f\|_q ,$$

where we have used the definition of T^* , Hölder's inequality and (3.6). From the inequality above we deduce that either $\lambda^+(y) = 0$ or $[\lambda^+(y)]^{1/q} \leqq (A_p/y)\cdot\|f\|_q$.

We have a similar estimate for $\lambda^-(y)$, and hence

$$\lambda(y) = \lambda^+(y) + \lambda^-(y) \leqq 2\left(\frac{A_p}{y}\right)^q \|f\|_q^q$$

which is (3.7) with $B_q = 2^{1/q}\cdot A_p$. □

Theorem 3.2. (Stein-Weiss). *If the operator T is linear and of restricted weak type p_o and p_1 (where $1 < p_o < p_1 < +\infty$), then T is of type p for every $p \in]p_o, p_1[$.*

Proof. Let $p \in]p_o, p_1[$ be given. From Lemma 1.8 we get that T is of restricted type p' for all $p' \in]p_o, p_1[$, and then lemma 3.1 tells us that T^* is of weak type q' for all $q' \in]q_1, q_o[$. We choose p_o' and p_1' such that $p_o < p_o' < p < p_1' < p_1$. Then T^* is of weak type q_1' and q_o', and from Marcinkiewicz' theorem (theorem 1.9) we infer that T^* is of type q, and especially that T^* is a bounded linear operator on L^q, which implies that $T^{**} = (T^*)^*$ is bounded on L^p. From (3.5) we conclude that $Tf = T^{**}f$ for all simple functions, i.e. on a dense set in L^p, and since both T and T^{**} are linear and T^{**} is of type p, the same is true for T. ◻

§ 4. Carleson-Hunt's theorem.

In this section we shall prove the main theorem under the assumption that an operator M defined below is of (strong) type p for all $p \in]1, +\infty[$. The rest of the text is devoted to the proof of this assertion.

We shall only consider functions f defined on the interval $[-\pi, \pi]$. We mention (without proofs) the following well-known facts about the spaces $L^p([-\pi, \pi])$.

(4.1) *If $1 \leq q \leq p \leq +\infty$, then $L^p([-\pi, \pi]) \subseteq L^q([-\pi, \pi])$*

(4.2) *Let $\varepsilon \in R_+$ be given. To any $f \in L^p([-\pi, \pi])$, $p \in [1, +\infty[$, one can find a polynomial P, such that $\|f - P\|_p < \varepsilon$.*

(4.3) *Let* $p \in [1, +\infty]$, *let* $f \in L^p([-\pi,\pi])$, *and let* (f_n) *be a sequence from* $L^p([-\pi,\pi])$, *such that* $\|f_n - f\|_p \to 0$ *as* $n \to +\infty$. *Then one can extract a subsequence* (f_{n_k}) *of* (f_n) , *such that*

$$f_{n_k}(x) \to f(x) \quad a.e.\ x \in [-\pi,\pi] \quad as \quad k \to +\infty .$$

From these properties it is easy to derive the following lemma.

<u>Lemma 4.1.</u> *Let* $p \in [1, +\infty[$, *and let* $f \in L^p([-\pi,\pi])$ *be a given function. To any given* $\varepsilon \in]0,1[$ *one can find a sequence* $(P_{\varepsilon,k})$ *of polynomials, such that*

(4.4) $$P_{\varepsilon,k}(x) \to f(x) \quad a.e.\ x \in [-\pi,\pi] \quad as \quad k \to +\infty$$

(4.5) $$\|f - P_{\varepsilon,k}\|_p^p < \varepsilon^{2k} \quad for\ all \quad k \in \mathbb{N}$$

<u>Proof.</u> Using (4.2) we can find a polynomial Q_k , such that $\|f - Q_k\|_p^p < \varepsilon^{2k}$

By (4.3) we can find an increasing sequence (n_k) from $\mathbb{N}$ (note that $n_k \geqq k$) , such that

$$P_{\varepsilon,k}(x) = Q_{n_k}(x) \to f(x) \quad a.e.\ x \in [-\pi,\pi] .$$

As ε^{2k} is decreasing in k and $n_k \geqq k$, the polynomials $(P_{\varepsilon,k})$ satisfy both (4.4) and (4.5). □

It follows from (4.1) that $L^p([-\pi,\pi]) \subseteq L^1([-\pi,\pi])$ for $p \in [1, +\infty]$. Let $f \in L^1([-\pi,\pi])$. Then the Fourier coefficients c_n , $n \in Z$, are all well-defined by

$$c_n = \frac{1}{2\pi} \int_{-\pi}^{\pi} f(t) e^{-int}\, dt , \quad n \in Z .$$

By $S_n(x;f)$ we denote the partial sum

$$S_n(x;f) = \sum_{k=-n}^{n} c_k e^{ikx} , \quad x \in [-\pi,\pi] , \quad n \in N_o ,$$

of the Fourier series for f .

In the following we shall only consider the case, where $p \in]1, +\infty[$. Let $\mathcal{F}$ denote the class of functions with values in $[0, +\infty]$. We define an operator $M : L^p([-\pi,\pi]) \to \mathcal{F}$ by

$$(4.6) \qquad Mf(x) = \sup\{|S_n(x;f)| \mid n \in N_0\} .$$

It is obvious that M is sublinear,

$$M(f+g) \leqq Mf + Mg .$$

The following theorem is a very deep result, which shall be proved in a later section.

<u>Theorem 4.2.</u> *The operator* M *is of type* p *for all* $p \in]1, +\infty[$ *. More precisely, to every* $p \in]1, +\infty[$ *there exists a constant* $C_p \in R_+$ *, such that*

$$(4.7) \qquad \|Mf\|_p \leqq C_p\|f\|_p \qquad \text{for all } f \in L^p([-\pi,\pi]) .$$

Here we shall take theorem 4.2 for granted. Then it is easy to prove the following lemma.

<u>Lemma 4.3.</u> *Let* $p \in]1, +\infty[$ *and let* $f \in L^p([-\pi,\pi])$ *be a given function. Let* C_p *be the constant in (4.7). For* $\varepsilon \in]0,1[$ *let* $P_{\varepsilon,k}$ *be the polynomials defined in lemma 4.1. For each* $k \in N$ *we define*

$$E_{\varepsilon,k} = \left\{x \in [-\pi,\pi] \mid M(f-P_{\varepsilon,k})(x) > \varepsilon^{k/p}\right\} .$$

Then $m(E_{\varepsilon,k}) < C_p^p \varepsilon^k$.

Proof. From lemma 4.1 we get $\|f-P_{\varepsilon,k}\|_p^p < \varepsilon^{2k}$, so

$$m(E_{\varepsilon,k}) = \left(\frac{1}{\varepsilon}\right)^k \cdot \varepsilon^k m(E_{\varepsilon,k}) = \left(\frac{1}{\varepsilon}\right)^k \int_{E_{\varepsilon,k}} \varepsilon^k\,dx \leqq \left(\frac{1}{\varepsilon}\right)^k \int_{E_{\varepsilon,k}} \{M(f-P_{\varepsilon,k})(x)\}^p dx$$

$$\leqq \left(\frac{1}{\varepsilon}\right)^k \int_{-\pi}^{\pi} \{M(f-P_{\varepsilon,k})(x)\}^p\,dx = \left(\frac{1}{\varepsilon}\right)^k \|M(f-P_{\varepsilon,k})\|_p^p \leqq \left(\frac{1}{\varepsilon}\right)^k C_p^p \|f-P_{\varepsilon,k}\|_p^p$$

$$< \left(\frac{1}{\varepsilon}\right)^k C_p^p \cdot \varepsilon^{2k} = C_p^p \varepsilon^k . \qquad \square$$

Theorem 4.4. (Carleson-Hunt, [1], [4], [7]). *Let* $p \in]1,+\infty]$. *For every* $f \in L^p([-\pi,\pi])$ *we have*

$$S_n(x;f) \to f(x) \qquad a.e.\ x \in [-\pi,\pi] \qquad as\ n \to +\infty .$$

Proof. As $L^\infty([-\pi,\pi]) \subseteqq L^p([-\pi,\pi])$ it is enough to consider the case $p \in]1,+\infty[$. Let $\varepsilon \in]0,1[$ be an arbitrary constant, and let $P_{\varepsilon,k}$ be the polynomials defined in lemma 4.1. As $S_n(x;f)$ is a partial sum of the Fourier series it is trivial that

$$S_n(x;f) = S_n(x;f-P_{\varepsilon,k}) + S_n(x;P_{\varepsilon,k})$$

for all $k \in N$ and all $x \in [-\pi,\pi]$. This implies for all $k, n \in N$ and all $x \in [-\pi,\pi]$ that

$$|S_n(x;f) - f(x)| \leqq |S_n(x;f-P_{\varepsilon,k})| + |S_n(x;P_{\varepsilon,k}) - f(x)| .$$

Since $P_{\varepsilon,k}(x)$ is a C^∞-function on $]-\pi,\pi[$, we have

$$\lim_{n\to+\infty} S_n(x;P_{\varepsilon,k}) = P_{\varepsilon,k}(x) \qquad \text{for all } x \in]-\pi,\pi[.$$

Hence, for every $k \in N$ and every $x \in]-\pi,\pi[$,

$$\limsup_{n\to+\infty} |S_n(x;f) - f(x)| \leqq \limsup_{n\to+\infty} |S_n(x;f-P_{\varepsilon,k})| + |f(x) - P_{\varepsilon,k}(x)|$$

Using that $\limsup_{n\to+\infty} |S_n(x;f-P_{\varepsilon,k})| \leqq M(f-P_{\varepsilon,k})(x)$ for all $k \in N$ and all $x \in]-\pi,\pi[$ we infer that

$$\limsup_{n\to+\infty} |S_n(x;f) - f(x)| \leqq M(f-P_{\varepsilon,k})(x) + |f(x) - P_{\varepsilon,k}(x)| . \tag{4.8}$$

Let $E_{\varepsilon,k}$ be defined as in lemma 4.3. Then $m(E_{\varepsilon,k}) \leqq C_p^p \varepsilon^k$. Furthermore, by the definition of $E_{\varepsilon,k}$,

$$M(f - P_{\varepsilon,k})(x) \leqq \varepsilon^{k/p} \qquad \text{for} \quad x \notin E_{\varepsilon,k} ,$$

so

$$M(f - P_{\varepsilon,k})(x) \to 0 \quad \text{for all} \quad x \notin A_\varepsilon = \bigcup_{n=1}^{+\infty} E_{\varepsilon,n} \quad \text{as} \quad k \to +\infty .$$

Hence, we only have to estimate the measure of the set $A_\varepsilon = \bigcup_{n=1}^{+\infty} E_{\varepsilon,n}$.
Since $\varepsilon \in]0,1[$ we get

$$m(A_\varepsilon) \leqq \sum_{n=1}^{+\infty} m(E_{\varepsilon,n}) \leqq C_p^p \sum_{n=1}^{+\infty} \varepsilon^n = C_p^p \cdot \frac{\varepsilon}{1-\varepsilon}$$

According to lemma 4.1 there exists a null-set B_ε , $m(B_\varepsilon) = 0$, such that

$$|f(x) - P_{\varepsilon,k}(x)| \to 0 \qquad \text{for} \quad x \in]-\pi,\pi[\setminus B_c .$$

Hence it follows from (4.8) that

$$\lim_{n\to+\infty} S_n(x;f) = f(x) \qquad \text{for all} \quad x \in]-\pi,\pi[\setminus(A_\varepsilon \cup B_\varepsilon) ,$$

where $m(A_\varepsilon \cup B_\varepsilon) \leqq C_p^p \cdot \frac{\varepsilon}{1-\varepsilon}$. As $\varepsilon \in]0,1[$ can be chosen arbitrarily small the theorem is proved. □

CHAPTER II.

In this chapter we shall define the Hilbert transform and the maximal Hilbert transform. These two transforms will play a crucial role later on. We shall follow [2]. The classical exposition may be found in [8].

The two transforms are formally defined in § 5 . It follows immediately from the definition that the maximal Hilbert transform is well-defined. This is not, however, so obvious in the case of the Hilbert transform itself. The proof of this assertion is postponed to § 6 . We use two auxiliary transforms P_y and Q_y combined with the Hardy-Littlewood maximal operator in order to derive the properties of the Hilbert transform.

As mentioned above we prove in § 6 that the Hilbert transform is well-defined. Furthermore we show that both the Hilbert transform and the maximal Hilbert transform are operators of type p for all $p \in]1,+\infty[$.

Finally, in § 7 , we prove some exponential estimates for the Hilbert transform and the maximal Hilbert transform. These results are similar to the exponential estimate for the Hardy-Littlewood maximal operator proved in theorem 2.6.

§ 5. The operators P_y and Q_y .

Let $f \in L^1(R)$. For each $y \in R_+$ let $H_y f$ be the function defined by

(5.1) $$H_y f(x) = \frac{1}{\pi} \int_{\{|x-t| \geq y\}} \frac{f(t)}{x-t}\, dt \, , \quad x \in R \, .$$

The *Hilbert transform* Hf of f is then defined as

(5.2) $$Hf(x) = \lim_{y \to 0+} H_y f(x) = \frac{1}{\pi} \, (pv) \int \frac{f(t)}{x-t}\, dt \, ,$$

where (pv) stands for "principal value". At this stage it is not clear, however, that Hf(x) is a well-defined expression. We shall in § 6 show that Hf(x) does exist almost everywhere.

We shall also consider the *maximal Hilbert transform* H^*f of f , which is defined by

(5.3) $$H^*f(x) = \sup\{|H_y f(x)| \mid y \in R_+\} \, .$$

In the applications we shall only consider functions f defined on a finite interval, so without loss of generality we may in the following assume that all functions f,g, ... under consideration have compact support. This assumption simplifies some of the proofs. It may be noted, however, that the results hold in general.

We introduce two linear operators P_y and Q_y related to H_y as follows

(5.4) $$P_y f(x) = \frac{1}{\pi} \int_{-\infty}^{+\infty} f(t) \cdot \frac{y}{(x-t)^2+y^2}\, dt \, , \quad y \in R_+ \, ,$$

(5.5) $$Q_y f(x) = \frac{1}{\pi} \int_{-\infty}^{+\infty} f(t) \cdot \frac{x-t}{(x-t)^2+y^2}\, dt \, , \quad y \in R_+ \, .$$

Clearly, $P_y f(x)$ and $Q_y f(x)$, $y \in R_+$, are well-defined for all $x \in R$, and *formally* we have $Q_o f(x) = Hf(x)$.

In the following y will always denote a positive number. We note that $P_y f(x)$ is the convolution of f with the integrable function $k_y(x) = \frac{1}{\pi} \cdot \frac{y}{x^2+y^2} = -\frac{1}{\pi} \operatorname{Im}\left(\frac{1}{z}\right)$, $z = x+iy$, i.e. $P_y f(x) = (f * k_y)(x)$, and $Q_y f(x)$ is similarly the convolution of f with the function $\ell_y(x) = \frac{1}{\pi} \cdot \frac{x}{x^2+y^2} = \frac{1}{\pi} \operatorname{Re}\left(\frac{1}{z}\right)$, i.e. $Q_y f(x) = (f * \ell_y)(x)$. However, ℓ_y is not integrable. Here and in the following, $z = x+iy$.

Concerning k_y we note that $k_y(x) > 0$ and $\int_{-\infty}^{+\infty} k_y(x)dx = 1$ and k_y is even, and decreasing on $[0,+\infty[$.

If $f(x)$ is a *continuous* L^∞-function, it is easy to see that $\varphi(x,y) = P_y f(x) = (f * k_y)(x)$ is the only *bounded* solution of the Dirichlet problem in the upper half plane for the Laplace operator

$$\text{(5.6)} \qquad \begin{cases} \Delta\varphi(x,y) = 0 & \text{for } x \in R ,\ y \in R_+ , \\ \varphi(x,0) = f(x) & \text{for } x \in R . \end{cases}$$

The formulæ (5.7), (5.8) and (5.9) below, which we shall need later on, may be proved in different ways. The most boring method uses a partial fraction expansion of the integrands and freshman's calculus. It is easier, though still tedious, to use residue calculus. A third method is to use that k_y in probability theory is the density function for a Cauchy distribution and that the Cauchy distribution is reproductive in the sense given below in (5.7). A fourth way is to use the remark above concerning the Dirichlet problem (5.6), guessing a bounded harmonic function $\varphi(x,y)$ in the upper half plane satisfying the boundary condition $\varphi(x,0) = f(x)$. Due to the uniqueness of the bounded solution of (5.6) this function $\varphi(x,y)$ is equal to the integral in (5.4). In the proofs below we shall select the shortest method.

First we prove

$$(5.7)\qquad \frac{1}{\pi}\int_{-\infty}^{+\infty} \frac{y_2}{(t-x_2)^2+y_2^2} \cdot \frac{y_1}{(x_1-t)^2+y_1^2}\,dt = \frac{y_1+y_2}{(x_1-x_2)^2+(y_1+y_2)^2} .$$

Let $z = x+iy$, $y \geqq 0$. Choose

$$\varphi(x,y) = -\operatorname{Im}\left(\frac{1}{z-x_2+iy_2}\right) = \frac{y+y_2}{(x-x_2)^2+(y+y_2)^2} .$$

Then $\varphi(x,y)$ is bounded and harmonic in the upper half plane, and $\varphi(x,0) = f(x) = \dfrac{y_2}{(x-x_2)^2+y_2^2}$ is bounded and continuous, so $\varphi(x,y)$ is the bounded solution of (5.6). Hence, $\varphi(x_1,y_1) = (f * k_{y_1})(x_1)$, which is exactly (5.7).

Similarly one proves

$$(5.8)\qquad \frac{1}{\pi}\int_{-\infty}^{+\infty} \frac{t-x_2}{(t-x_2)^2+y_2^2} \cdot \frac{y_1}{(x_1-t)^2+y_1^2}\,dt = \frac{x_1-x_2}{(x_1-x_2)^2+(y_1+y_2)^2} .$$

In this case $\varphi(x,y)$ is chosen as

$$\varphi(x,y) = \operatorname{Re}\left(\frac{1}{z-x_2+iy_2}\right) = \frac{x-x_2}{(x-x_2)^2+(y+y_2)^2} .$$

Now, consider the function $f(z) = \dfrac{1}{x_1-iy_1-z} \cdot \dfrac{1}{z-x_2-iy_2}$, which is analytic in the upper half plane except for the simple pole x_2+iy_2 . Using residue calculus (cf. e.g. [6]) we get

$$\int_{-\infty}^{+\infty} \frac{1}{x_1-iy_1-t} \cdot \frac{1}{t-x_2-iy_2}\,dt = 2\pi i \operatorname{Res}\left[\frac{1}{x_1-iy_1-z} \cdot \frac{1}{z-x_2-iy_2} ; x_2+iy_2\right]$$

$$= \frac{2\pi i}{x_1-x_2-i(y_1+y_2)} .$$

Taking the real part of this equation we get

$$\int_{-\infty}^{+\infty} \frac{x_1-t}{(x_1-t)^2+y_1^2} \cdot \frac{t-x_2}{(t-x_2)^2+y_2^2}\, dt - \int_{-\infty}^{+\infty} \frac{y_1}{(x_1-t)^2+y_1^2} \cdot \frac{y_2}{(t-x_2)^2+y_2^2}\, dt$$

$$= - 2\pi \frac{y_1+y_2}{(x_1-x_2)^2+(y_1+y_2)^2} .$$

Comparing this result with (5.7) we finally get the formula

$$\text{(5.9)} \qquad \frac{1}{\pi} \int_{-\infty}^{+\infty} \frac{t-x_1}{(x_1-t)^2+y_1^2} \cdot \frac{t-x_2}{(t-x_2)^2+y_2^2}\, dt = \frac{y_1+y_2}{(x_1-x_2)^2+(y_1+y_2)^2} .$$

We shall now prove a few results concerning the operators P_y and Q_y .

Lemma 5.1. *For every integrable function we have*

$$\text{(5.10)} \qquad Q_{y_1}(Q_{y_2} f) = -P_{y_1+y_2} f ,$$

$$\text{(5.11)} \qquad P_{y_1}(Q_{y_2} f) = Q_{y_1+y_2} f .$$

Proof. Applying Fubini's theorem and (5.9) we have the following computation

$$Q_{y_1}(Q_{y_2} f)(x) = \frac{1}{\pi} \int_{-\infty}^{+\infty} Q_{y_2} f(t) \cdot \frac{x-t}{(x-t)^2+y_1^2}\, dt$$

$$= \frac{1}{\pi} \int_{t=-\infty}^{+\infty} \left\{ \frac{1}{\pi} \int_{u=-\infty}^{+\infty} f(u) \cdot \frac{t-u}{(t-u)^2+y_2^2}\, du \right\} \frac{x-t}{(x-t)^2+y_1^2}\, dt$$

$$= \frac{1}{\pi} \int_{u=-\infty}^{+\infty} f(u) \left\{ \frac{1}{\pi} \int_{t=-\infty}^{+\infty} \frac{x-t}{(x-t)^2+y_1^2} \cdot \frac{t-u}{(t-u)^2+y_2^2}\, dt \right\} du$$

$$= \frac{1}{\pi} \int_{u=-\infty}^{+\infty} f(u) \cdot \frac{-(y_1+y_2)}{(x-u)^2+(y_1+y_2)^2}\, du = -P_{y_1+y_2} f(x) .$$

By another application of Fubini's theorem and (5.8) we have the computation

$$P_{y_1}(Q_{y_2} f)(x) = \frac{1}{\pi}\int_{-\infty}^{+\infty} Q_{y_2} f(t) \frac{y_1}{(x-t)^2+y_1^2}\, dt$$

$$= \frac{1}{\pi}\int_{t=-\infty}^{+\infty} \left\{\frac{1}{\pi}\int_{u=-\infty}^{+\infty} f(u) \frac{t-u}{(t-u)^2+y_2^2}\, du\right\} \frac{y_1}{(x-t)^2+y_1^2}\, dt$$

$$= \frac{1}{\pi}\int_{u=-\infty}^{+\infty} f(u) \left\{\frac{1}{\pi}\int_{t=-\infty}^{+\infty} \frac{t-u}{(t-u)^2+y_2^2} \cdot \frac{y_1}{(x-t)^2+y_1^2}\, dt\right\} du$$

$$= \frac{1}{\pi}\int_{u=-\infty}^{+\infty} f(u) \cdot \frac{x-u}{(x-u)^2+(y_1+y_2)^2}\, du = Q_{y_1+y_2} f(x) \ . \qquad \square$$

Lemma 5.2. *For every* $f, g \in L^2(R)$ *we have*

$$\int_{-\infty}^{+\infty} f \cdot Q_y g\, dx = -\int_{-\infty}^{+\infty} Q_y f \cdot g\, dx \tag{5.12}$$

$$\int_{-\infty}^{+\infty} \left\{Q_{y_1+y_2} f - Q_{y_1} f\right\}^2 dx = \int_{-\infty}^{+\infty} f \cdot \left\{P_{2y_1+2y_2} f - 2P_{2y_1+y_2} f + P_{2y_1} f\right\} dx \ . \tag{5.13}$$

Proof. By Fubini's theorem we have

$$\int_{-\infty}^{+\infty} f(x)\, Q_y g(x) dx = \int_{x=-\infty}^{+\infty} f(x) \left\{\frac{1}{\pi}\int_{t=-\infty}^{+\infty} g(t) \cdot \frac{x-t}{(x-t)^2+y^2} dt\right\} dx$$

$$= -\int_{t=-\infty}^{+\infty} g(t) \left\{\frac{1}{\pi}\int_{x=-\infty}^{+\infty} f(x) \cdot \frac{t-x}{(t-x)^2+y^2} dx\right\} dt = -\int_{-\infty}^{+\infty} g(t)\, Q_y f(t) dt \ .$$

To prove (5.13) we compute

$$\int_{-\infty}^{+\infty} Q_{y_1+y_2} f \cdot Q_{y_1+y_2} f\, dx \tag{5.14}$$

$$= \int_{-\infty}^{+\infty} \left\{\frac{1}{\pi}\int_{-\infty}^{+\infty} f(t) \frac{x-t}{(x-t)^2+(y_1+y_2)^2} dt\right\} \left\{\frac{1}{\pi}\int_{-\infty}^{+\infty} f(u) \frac{x-u}{(x-u)^2+(y_1+y_2)^2} du\right\} dx$$

$$= \int_{t=-\infty}^{+\infty} f(t) \left\{\frac{1}{\pi}\int_{u=-\infty}^{+\infty} f(u) \left(\frac{1}{\pi}\int_{x=-\infty}^{+\infty} \frac{x-t}{(x-t)^2+(y_1+y_2)^2} \cdot \frac{x-u}{(x-u)^2+(y_1+y_2)^2}\, dx\right) du\right\} dt$$

$$= \int_{t=-\infty}^{+\infty} f(t) \left\{ \frac{1}{\pi} \int_{u=-\infty}^{+\infty} f(u) \frac{2y_1+2y_2}{(t-u)^2+(2y_1+2y_2)^2} du \right\} dt = \int_{-\infty}^{+\infty} f(t) P_{2y_1+2y_2} f(t) dt ,$$

where we again have used Fubini's theorem and the identity (5.9). The other terms are handled in the same way. □

Our next aim will be to relate the operator P_y to the Hardy-Littlewood maximal operator Θ . We have

$$P_y f(x) = (f*k_y)(x) = \int_{-\infty}^{+\infty} f(x-t) k_y(t) dt = \int_{-\infty}^{+\infty} f(x-t) \left\{ \int_{u=0}^{k_y(t)} du \right\} dt$$

$$= \int_{u=0}^{+\infty} \left\{ \int_{\{|t| \leqq k_y^{-1}(u)\}} f(x-t) dt \right\} du$$

$$= \int_{u=0}^{+\infty} 2k_y^{-1}(u) \left\{ \frac{1}{2k_y^{-1}(u)} \int_{\{|t| \leqq k_y^{-1}(u)\}} f(x-t) dt \right\} du$$

$$\leqq \int_0^{+\infty} 2k_y^{-1}(u)\, \Theta f(x) du = \Theta f(x) \quad ,$$

where Θ is the Hardy-Littlewood maximal operator introduced in § 2 , and k_y^{-1} is the inverse function of the restriction of k_y to $[0,+\infty[$. Thus

(5.15) $$|P_y f(x)| \leqq (P_y|f|)(x) \leqq \Theta f(x) \qquad [=(\Theta|f|)(x)]$$

for all $x \in R$ and for all $y \in R_+$.

It was earlier remarked that *formally* $Q_0 f(x) = Hf(x)$. More precisely we have

Lemma 5.3. *For every integrable function* f

(5.16) $$\lim_{y\to 0+} \{H_y f(x) - Q_y f(x)\} = 0 \quad a.e.$$

Proof. We have

$$H_y f(x) - Q_y f(x) = \frac{1}{\pi} \int_{\{|x-t| \geqq y\}} f(t) \left[\frac{1}{x-t} - \frac{x-t}{(x-t)^2+y^2} \right] dt$$

$$+ \frac{1}{\pi} \int_{\{|x-t|<y\}} f(t) \cdot \frac{x-t}{(x-t)^2+y^2}\, dt ,$$

from which we get (using that $|x-t| \geqq y$ in the former integral and that $|x-t| < y$ in the latter one)

$$|H_y f(x) - Q_y f(x)| \leqq \frac{1}{\pi} \int_{\{|x-t| \geqq y\}} \frac{|f(t)|y^2}{|x-t|\{(x-t)^2+y^2\}}\, dt + \frac{1}{\pi} \int_{\{|x-t|<y\}} \frac{|f(t)|y}{(x-t)^2+y^2}\, dt$$

$$\leqq \frac{1}{\pi} \left(\int_{\{|x-t| \geqq y\}} + \int_{\{|x-t|<y\}} \right) \frac{|f(t)|y}{(x-t)^2+y^2}\, dt = P_y |f|(x) ,$$

so we have proved that

$$(5.17) \qquad |H_y f(x) - Q_y f(x)| \leqq (P_y |f|)(x) .$$

From the integral representation of $H_y f - Q_y f$ above it follows that the operator $H_y - Q_y$ may be extended to all constants a and that $(H_y - Q_y)a = 0$. Thus, the left hand side of (5.17) remains unaffected, if we replace f by $f - a$. Choosing $a = f(x)$ we see that we in the right hand side of (5.17) can replace $f(t)$ by $f(t) - f(x)$, giving

$$(5.18) \qquad |H_y f(x) - Q_y f(x)| \leqq \frac{1}{\pi} \int_{-\infty}^{+\infty} |f(t) - f(x)| \cdot \frac{y}{(x-t)^2+y^2}\, dt .$$

Thus (5.16) follows, if we can prove that

$$(5.19) \qquad \lim_{y \to 0+} \frac{1}{\pi} \int_{-\infty}^{+\infty} |f(t) - f(x)| \cdot \frac{y}{(x-t)^2+y^2}\, dt = 0 \quad \text{for a.e. } x \in R .$$

Clearly, (5.19) holds if f is continuous. To prove (5.19) in general we put

$$\Omega f(x) = \limsup_{y \to 0+} |H_y f(x) - Q_y f(x)| .$$

From (5.17) and (5.15) follows that $\Omega f(x) \leqq \Theta f(x)$, so from theorem 2.1 in § 2 we get

$$\text{(5.20)} \qquad m(\{x \mid \Omega f(x) > y\}) \leqq \frac{4}{y}\int_{-\infty}^{+\infty} |f(x)|dx \ .$$

Now, the left hand side of (5.20) is unaffected, if we from f subtract any continuous function f_o of compact support. Thus,

$$m(\{x \mid \Omega f(x) > y\}) \leqq \frac{4}{y}\int_{-\infty}^{+\infty} |f(x) - f_o(x)|dx \ .$$

Using the fact that the class of continuous functions of compact support is dense in $L^1(R)$ we deduce that $m(\{x \mid \Omega f(x) > y\}) = 0$ for all $y > 0$, and so $\Omega f(x) = 0$ a.e., proving the lemma. □

We have the following result concerning P_y.

__Theorem 5.4.__ *If $f \in L^p$ for some $p \in]1,+\infty]$, then also $P_y f \in L^p$, and*

$$\text{(5.21)} \qquad \|P_y f\|_p \leqq \|f\|_p \ ,$$

i.e. the operator P_y is of strong type p for all $p \in]1,+\infty]$.

__Proof.__ Using Hölder's inequality we get for $p \in]1,+\infty[$

$$|P_y f(x)|^p \leqq \left(\int_{-\infty}^{+\infty} k_y(x-t)\cdot |f(t)|dt\right)^p = \left\{\int_{-\infty}^{+\infty} (k_y(x-t)^{1/p}\cdot |f(t)|)\cdot k_y(x-t)^{1/q}dt\right\}^p$$

$$\leqq \int_{-\infty}^{+\infty} k_y(x-t)|f(t)|^p dt \cdot \left(\int_{-\infty}^{+\infty} k_y(x-t)dt\right)^{p/q} = \int_{-\infty}^{+\infty} k_y(x-t)|f(t)|^p dt \ .$$

Integration with respect to x gives

$$\|P_y f\|_p^p \leqq \int_{-\infty}^{+\infty} |f(t)|^p\left\{\int_{-\infty}^{+\infty} k_y(x-t)dx\right\}dt = \|f\|_p^p \ .$$

For $p = +\infty$ we get

$$|P_y f(x)| \leqq \int_{-\infty}^{+\infty} k_y(x-t)\cdot |f(t)|dt \leqq \|f\|_\infty \int_{-\infty}^{+\infty} k_y(x-t)dt = \|f\|_\infty \ . \qquad \square$$

Remark 5.5. We may get a shorter proof from (5.15), if we apply theorem 2.1 and corollary 2.2. In fact,

$$\|P_y f\|_p \leqq \|\Theta f\|_p \leqq C_p \|f\|_p \quad , \quad p \in]1,+\infty] \ .$$

However, we do not get that the constant may be chosen equal 1 .

As a corollary we have

Corollary 5.6. *If* $f \in L^p$ *for some* $p \in]1,+\infty[$ *, then*

$$\lim_{y\to 0+} \|P_y f - f\|_p = 0 \tag{5.22}$$

and

$$\lim_{y\to 0+} P_y f(x) = f(x) \quad \text{for almost every } x \in \mathbb{R} \ . \tag{5.23}$$

Proof. First we notice that (5.22) and (5.23) are obvious if f is a continuous function of compact support. Let $f \in L^p$ and let $\varepsilon \in \mathbb{R}_+$ be given. Then there exists a continuous function f_ε of compact support such that $\|f - f_\varepsilon\|_p < \varepsilon$, so we have by Minkowski's inequality

$$\begin{aligned} \|P_y f - f\|_p &\leqq \|P_y(f - f_\varepsilon)\|_p + \|P_y f_\varepsilon - f_\varepsilon\|_p + \|f_\varepsilon - f\|_p \\ &\leqq 2\|f - f_\varepsilon\|_p + \|P_y f_\varepsilon - f_\varepsilon\|_p \ , \end{aligned}$$

and thus

$$\limsup_{y\to 0+} \|P_y f - f\|_p \leqq 2\varepsilon \quad \text{for all} \quad \varepsilon > 0 \ ,$$

hence (5.22) holds.

To prove (5.23) we let

$$\Omega f(x) = \limsup_{y\to 0+} |P_y f(x) - f(x)| \ .$$

We note that

$$\Omega f(x) \leqq \Theta f(x) + |f(x)| \ .$$

Now the proof follows the same line as the proof of (5.19). □

Unfortunately, we cannot in the same way as in theorem 5.4 get estimates for $\|Q_y f\|_p$, because the function ℓ_y is not integrable. Neither is it possible in that way to get an estimate for $\|H_y f\|_p$.

§ 6. Existence of the Hilbert transform and estimates for the Hilbert transform and the maximal Hilbert transform.

The purpose of this section is to show that the limit

$$(6.1) \qquad Hf(x) = \lim_{y \to 0+} \frac{1}{\pi} \int_{\{|x-t| \geq y\}} \frac{f(t)}{x-t}\, dt$$

exists almost everywhere for f integrable and to prove estimates for both the Hilbert transform H and the maximal Hilbert transform H^* , defined by (5.3).

We start with a lemma due to Loomis [5] .

Lemma 6.1. *Let* $c_1, c_2, \ldots, c_n$ *be positive constants and let* $x_1 < x_2 < \ldots < x_n$ *be given real numbers. We define a function* g *by*

$$g(x) = \sum_{j=1}^{n} \frac{c_j}{x-x_j} .$$

Then for each $\lambda \in R_+$

$$(6.2) \qquad m(\{x \mid g(x) > \lambda\}) = m(\{x \mid g(x) < -\lambda\}) = \frac{1}{\lambda} \sum_{j=1}^{n} c_j .$$

Proof. We shall in detail prove the statement for $m(\{x \mid g(x) > \lambda\})$. With trivial modifications the proof for $m(\{x \mid g(x) < -\lambda\})$ follows the same line.

In each of the intervals $]x_j, x_{j+1}[$, $j = 1,2, \ldots, n-1$, the function $g(x)$ decreases from $+\infty$ to $-\infty$. In $]-\infty, x_1[$ the function $g(x)$ decreases from 0 to $-\infty$, and in $]x_n, +\infty[$ it decreases from $+\infty$ to 0 . Thus, if $a_1(\lambda), \ldots, a_n(\lambda)$ are the roots of the equation

$$(6.3) \qquad g(x) = \lambda ,$$

where $a_j(\lambda) \in]x_j, x_{j+1}[$, $j = 1,2, \ldots, n-1$, and $a_n(\lambda) \in]x_n, +\infty[$, we must have

(6.4) $$m(\{x \mid g(x) > \lambda\}) = \sum_{j=1}^{n} \{a_j(\lambda) - x_j\} .$$

Multiplying both sides of (6.3) by $\frac{1}{\lambda} \prod_{j=1}^{n} (x - x_j)$ we get that $a_1(\lambda), \ldots, a_n(\lambda)$ are the roots of the normalized polynomial

$$\prod_{j=1}^{n} (x - x_j) - \frac{1}{\lambda} \sum_{j=1}^{n} c_j \prod_{k=1}^{n}{}^{(j)} (x - x_k) = 0 ,$$

where $\prod_{k=1}^{n}{}^{(j)} (x - x_k) = (x - x_1) \ldots (x - x_{j-1})(x - x_{j+1}) \ldots (x - x_n)$. As the sum of the roots is equal to the coefficient of x^{n-1} with the opposite sign we finally get

$$\sum_{j=1}^{n} a_j(\lambda) = \sum_{j=1}^{n} x_j + \frac{1}{\lambda} \sum_{j=1}^{n} c_j \quad \text{or} \quad \sum_{j=1}^{n} \{a_j(\lambda) - x_j\} = \frac{1}{\lambda} \sum_{j=1}^{n} c_j .$$

Substituting the latter relation into (6.4) we get (6.2). □

Theorem 6.2. *The operator* H^* *is of weak type 1.*

Proof. We shall prove that

(6.5) $$m(\{x \mid H^*f(x) > \lambda\}) \leq \frac{128}{\lambda\pi} \|f\|_1 ,$$

for all $\lambda \in \mathbb{R}_+$ and all $f \in L^1(\mathbb{R})$ with compact support (cf. a remark in § 2 and definition 1.4).

In general, $f = f^+ - f^-$, where $f^+, f^- \geq 0$ and $|f| = f^+ + f^-$. Furthermore,

(6.6) $$m(\{x \mid H^*f(x) > \lambda\}) \leq m(\{x \mid H^*(f^+)(x) > \tfrac{\lambda}{2}\}) + m(\{x \mid H^*(f^-)(x) > \tfrac{\lambda}{2}\}) ,$$

so we may in the following assume that $f \geq 0$.

Let $\lambda \in \mathbb{R}_+$ be given. For each $\varepsilon \in \mathbb{R}_+$ we define the two sets

$$E_\varepsilon^+ = \{x \mid \sup_{y \geq \varepsilon} H_y f(x) > \lambda\} , \qquad E_\varepsilon^- = \{x \mid \sup_{y \geq \varepsilon} (-H_y f(x)) > \lambda\}$$

[Note that even if f is non-negative, $H_y f$ need not be non-negative; cf. (5.1).] We shall only consider the set E_ε^+ and prove that $m(E_\varepsilon^+) \leq \frac{32}{\lambda\pi} \int_{-\infty}^{+\infty} f(t)dt$, as the proof for E_ε^- is analogous.

For any finite interval I of the real axis let $c(I)$ denote the center of I, and I^c the complement of I.

Let us consider the family of open intervals I, for which

(6.7) $$\frac{1}{\pi}\int_{I^c} f(t)\cdot\frac{1}{c(I)-t}dt > \lambda .$$

By the definition of E_ε^+ these intervals I cover E_ε^+, and as E_ε^+ is bounded (here we use that f has compact support) already a finite number of these intervals cover E_ε^+. Using the same reasoning of Besicovitch type as in the proof of theorem 2.1 we can find disjoint intervals $I_1, \dots, I_n$ such that

(6.8) $$m(E_\varepsilon^+) \leqq 4 \sum_{j=1}^{n} m(I_j)$$

and

(6.9) $$\frac{1}{\pi}\int_{I_j^c} f(t)\cdot\frac{1}{c(I_j)-t}dt > \lambda , \quad j=1,2,\dots,n .$$

The function $g_x(t) = \frac{1}{x-t}$ is uniformly continuous for $|x-t| \geqq \varepsilon$, and it tends to 0 as $|t| \to +\infty$. Thus, to any $\delta \in]0,1[$ we can find a decomposition of the real axis into a finite number of small intervals J and two infinite intervals such that for each of the small intervals J and for each $j=1,2,\dots,n$,

(6.10) $$\left|\frac{1}{\pi}\int_{I_j^c} f(t)\cdot\frac{1}{c(I_j)-t}dt - \frac{1}{\pi}\sum_{J\cap I_j=\emptyset}\int_J f(t)dt\cdot\frac{1}{c(I_j)-c(J)}\right| < \delta\cdot\lambda .$$

We can even choose the intervals J in such a way that for each I_j either $J \subseteqq I_j$ or $J \cap I_j = \emptyset$. In the following we shall suppose that this has been done. Due to the facts that f has compact support and $g_x(t)$ tends to 0 as $|t| \to +\infty$ we shall never need to consider the two infinite intervals mentioned above.

We define

$$g(x) = \frac{1}{\pi}\sum_J \int_J f(t)dt\cdot\frac{1}{x-c(J)} \quad \text{and} \quad g_j(x) = \frac{1}{\pi}\sum_{J\subseteqq I_j}\int_J f(t)dt\cdot\frac{1}{x-c(J)} .$$

The function

$$g(x) - g_j(x) = \frac{1}{\pi} \sum_{J \cap I_j = \emptyset} \int_J f(t)dt \cdot \frac{1}{x - c(J)}$$

is clearly decreasing in the interval I_j , and from (6.9) and (6.10) we get for $x = c(I_j)$,

$$g(c(I_j)) - g_j(c(I_j)) > (1-\delta)\lambda \ , \quad \text{so} \quad g(x) - g_j(x) > (1-\delta)\lambda$$

for x in the left half of I_j . Thus we deduce that

$$\sum_{j=1}^{n} \frac{1}{2} m(I_j) \leqq m(\{x \mid g(x) > \frac{\lambda}{2}(1-\delta)\}) + \sum_{j=1}^{n} m(\{x \mid g_j(x) < -\frac{\lambda}{2}(1-\delta)\}) \ .$$

An application of lemma 6.1 then gives

$$\sum_{j=1}^{n} \frac{1}{2} m(I_j) \leqq \frac{2}{\lambda(1-\delta)} \sum_J \frac{1}{\pi} \int_J f(t)dt + \sum_{j=1}^{n} \frac{2}{\lambda(1-\delta)} \sum_{J \subseteqq I_j} \frac{1}{\pi} \int_J f(t)dt \leqq \frac{4}{\lambda(1-\delta)} \cdot \frac{1}{\pi} \int_{-\infty}^{+\infty} f(t)dt.$$

Combining this inequality with (6.8) we get, letting $\delta \to 0+$,

$$m(E_\varepsilon^+) \leqq \frac{32}{\lambda\pi} \int_{-\infty}^{+\infty} f(t)dt$$

as mentioned above. In a similar way we get

$$m(E_\varepsilon^-) \leqq \frac{32}{\lambda\pi} \int_{-\infty}^{+\infty} f(t)dt \ .$$

Letting $\varepsilon \to 0+$ we get (for $f \geqq 0$)

$$m(\{x \mid H^*f(x) > \lambda\}) = m(\{x \mid \sup_{y \in R_+} |H_y f(x)| > \lambda\}) \leqq \frac{64}{\lambda\pi} \int_{-\infty}^{+\infty} f(t)dt \ .$$

Finally, we get for $f = f^+ - f^-$, using (6.6) ,

$$m(\{x \mid H^*f(x) > \lambda\}) \leqq \frac{128}{\lambda\pi} \int_{-\infty}^{+\infty} f^+(t)dt + \frac{128}{\lambda\pi} \int_{-\infty}^{+\infty} f^-(t)dt = \frac{128}{\lambda\pi} \|f\|_1 \ ,$$

which is (6.5), and we have proved that H^* is of weak type 1 . □

We shall now establish that

(6.11) $$H f(x) = \lim_{y \to 0+} H_y f(x)$$

exists almost everywhere. If the function f is continuously differentiable and of compact support, it follows from

$$H_y f(t) = \frac{1}{\pi}\int_y^{+\infty} \frac{f(x-t)+f(x+t)}{t}\,dt$$

that the limit in (6.11) exists for all x. For a general $f \in L^1(R)$ we set

$$\Omega f(x) = \limsup_{y_1\to 0+,\, y_2\to 0+} |H_{y_1} f(x) - H_{y_2} f(x)|$$

and obtain from theorem 6.2 that

$$m(\{x \mid \Omega f(x) > y\}) \leq \frac{C}{y}\|f\|_1 .$$

Now, $\Omega f(x)$ does not change if we from f subtract an arbitrary continuously differentiable function of compact support. Hence we deduce that

$$m(\{x \mid \Omega f(x) > y\}) = 0 \quad \text{for all} \quad y > 0 ,$$

and so $\Omega f(x) = 0$ almost everywhere. This means that the limit in (6.11) exists almost everywhere, and so we have established the existence of the Hilbert transform, which clearly is a linear operator.

Using lemma 5.3 we note that we also have

(6.12) $$H f(x) = \lim_{y\to 0+} Q_y f(x) \quad \text{a.e.}$$

Theorem 6.3. *If $f \in L^2(R)$, then $H f \in L^2(R)$ and*

(6.13) $$\lim_{y\to 0+} \|Hf - Q_y f\|_2 = 0 ,$$

(6.14) $$\lim_{y\to 0+} \|Hf - H_y f\|_2 = 0 ,$$

(6.15) $$\|Hf\|_2 = \|f\|_2 ,$$

(6.16) $$H(Hf) = -f , \quad \text{a.e.}$$

Proof. We start by proving (6.13). If we let $y_1 \to 0+$ in (5.13) we get

$$\int_{-\infty}^{+\infty} \{Q_{y_2} f - H f\}^2 dx \leq \int_{-\infty}^{+\infty} |f| \cdot |P_{2y_2} f - 2P_{y_2} f + f|\,dx ,$$

where we have used Fatou's lemma, (6.12) and corollary 5.6. Another application of corollary 5.6 gives (6.13).

We next prove (6.14). From lemma 5.3 we have $\lim_{y\to 0+}[H_y f(x) - Q_y f(x)] = 0$ almost everywhere. Furthermore, it follows from (5.17), (5.15) and corollary 2.2 that

$$\|H_y f - Q_y f\|_2^2 \leq \|\Theta f\|_2^2 \leq 32\|f\|_2^2 < +\infty ,$$

and thus $\|H_y f - Q_y f\|_2 \to 0$ as $y \to 0+$ by dominated convergence. Together with (6.13) this implies (6.14).

From (5.14) follows that

$$\|Q_y f\|_2^2 = \int |Q_y f|^2 dx = \int f \cdot P_{2y} f \, dx , \tag{6.17}$$

so an application of corollary 5.6 and (6.13) gives (6.15).

At last we prove (6.16). Using Minkowski's inequality we get

$$\|H(Hf)+f\|_2 \leq \|H(Hf)-Q_y(Hf)\|_2 + \|Q_y(Hf)-Q_y(Q_y f)\| + \|Q_y(Q_y f)+f\|_2 . \tag{6.18}$$

From theorem 5.4 and (6.17) we deduce

$$\|Q_y f\|_2^2 = \int_{-\infty}^{+\infty} f(x)\, P_{2y} f(x) dx \leq \|f\|_2 \cdot \|P_{2y} f\|_2 \leq \|f\|_2^2 .$$

Using the inequality and (5.10) in (6.18) we get

$$\|H(Hf) + f\|_2 \leq \|H(Hf) - Q_y(Hf)\|_2 + \|H f - Q_y f\|_2 + \|-P_{2y} f + f\|_2 .$$

Using $\|P_{2y} f - f\| \to 0$ [corollary 5.6] and $\|Q_y f - H f\| \to 0$ [(6.13) proved above] for $y \to 0+$ we infer that $\|H(Hf) + f\|_2 = 0$ proving (6.16). □

<u>Theorem 6.4.</u> *If* $f, g \in L^2(\mathbb{R})$, *then*

$$\int_{-\infty}^{+\infty} f \cdot Hg \, dx = -\int_{-\infty}^{+\infty} g \cdot Hf \, dx . \tag{6.19}$$

Proof. We start from the identity (5.12), let $y \to 0+$ and use (6.13). □

It follows from theorem 6.2 that H is of weak type 1. From (6.15) in theorem 6.3 we have that H is of type 2, and hence also of weak type 2. The Marcinkiewicz theorem (theorem 1.9) then gives that H is of type p for all $p \in]1,2[$, hence for all $p \in]1,2]$, i.e. there exists a constant $c_p \in R_+$, such that

$$(6.20) \qquad \|Hf\|_p \leq c_p \|f\|_p \quad \text{for all } f \in L^p(R) \quad (1 < p \leq 2) .$$

If $f \in L^p(R)$, $1 < p \leq 2$, and $g \in L^q(R)$, the identity (5.12) is still valid, so letting $y \to 0+$ and using that $|\int g \cdot Hf\,dx| \leq c_p \|f\|_p \|g\|_q$ we conclude that (6.19) also holds for $f \in L^p(R)$ and $g \in L^q(R)$. Especially we get

$$(6.21) \qquad \left|\int_{-\infty}^{+\infty} f \cdot Hg\,dx\right| = \left|\int_{-\infty}^{+\infty} g \cdot Hf\,dx\right| \leq c_p \|f\|_p \|g\|_q , \quad 1 < p \leq 2 ,$$

for all $g \in L^q(R)$ and all $f \in L^p(R)$, so

$$\|Hg\|_q \leq \sup_{\|f\|_p \leq 1} \left|\int_{-\infty}^{+\infty} f \cdot Hg\,dx\right| \leq c_p \|g\|_q ,$$

proving that H is of type $q \in [2,+\infty[$, where $c_q = c_p$, $\frac{1}{p} + \frac{1}{q} = 1$. Thus we have proved

Theorem 6.5. *The operator H is of type p for all $p \in]1,+\infty[$.*

We shall at last in this section prove that the maximal Hilbert transform H^* also is of type p for $p \in]1,+\infty[$. First we prove

Lemma 6.6. *For $f \in L^p(R)$, $p \in]1,+\infty[$, we have*

$$H^*f(x) \leq \Theta f(x) + \Theta(Hf)(x) ,$$

where Θ is the Hardy-Littlewood maximal operator.

Proof. From (5.11) in lemma 5.1 we get

$$(6.22) \qquad P_y(Q_{y_2} f)(x) = Q_{y+y_2} f(x) .$$

When $f \in L^2(R)$ we deduce from (6.13) in theorem 6.3, letting $y_2 \to 0$ in (6.22) that

$$Q_y f(x) = P_y Hf(x) .$$

Using (5.17) we have

$$\begin{aligned} |H_y f(x)| &\leqq |H_y f(x) - Q_y f(x)| + |Q_y f(x)| \\ &= |H_y f(x) - Q_y f(x)| + |P_y Hf(x)| \\ &\leqq P_y(|f|)(x) + P_y(|Hf|)(x) . \end{aligned}$$

By an application of (5.15) we then get the result in lemma 6.6, first for all $f \in L^2(R) \cap L^p(R)$, and then for all $f \in L^p(R)$. □

As both H and Θ are of type p for all $p \in]1,+\infty[$ it easily follows from lemma 6.6 that we have

Theorem 6.7. *The maximal Hilbert transform* H^* *is of type* p *for all* $p \in]1,+\infty[$.

§ 7. Exponential estimates for the Hilbert transform and the maximal Hilbert transform.

We shall now prove some exponential estimates similar to the result given in theorem 2.6 for the Hardy-Littlewood maximal operator. We shall only consider functions, which are essentially bounded and which are equal to zero outside a finite interval of length A .

Theorem 7.1. *There exist positive constants* c_1 *and* c_2 *such that, if* f *is any essentially bounded function, the support of which is contained in an interval of length* A *, then*

$$(7.1) \qquad m(\{x \mid |Hf(x)| > \lambda\}) \leqq c_1 \cdot A \cdot \frac{\|f\|_\infty}{\lambda} \cdot \exp\left(-c_2 \cdot \frac{\lambda}{\|f\|_\infty}\right) .$$

Proof. We may assume that $\|f\|_\infty = 1$. Since H is of weak type 1 [using theorem 6.2 we have e.g. $m(\{x \mid |Hf(x)| > \lambda\}) \leqq \frac{128}{\lambda\pi} \|f\|_1$] it is enough to prove (7.1) for large values of λ . Let

$$E_\lambda^+ = \{x \mid H f(x) > \lambda\} , \qquad E_\lambda^- = \{x \mid H f(x) < -\lambda\} .$$

We shall give an estimate for $m(E_\lambda^+)$, the estimate for $m(E_\lambda^-)$ being similar.

We have, applying theorem 6.4,

$$(7.2) \qquad \lambda\, m(E_\lambda^+) \leq \int_{-\infty}^{+\infty} \chi_{E_\lambda^+}(x) \cdot H f(x)dx = -\int_{-\infty}^{+\infty} f(x) \cdot H \chi_{E_\lambda^+}(x)dx .$$

As H is of weak type 1 and of type 2 (according to theorem 6.3 we have $\|H f\|_2 = \|f\|_2$) we conclude from the Marcinkiewicz theorem (theorem 1.9) that for $p \in]1,2[$

$$\|H \chi_{E_\lambda^+}\|_p^p \leq 2^p \left(\frac{p}{p-1} + \frac{p}{2-p}\right) \left(\frac{128}{\pi}\right)^{2-p} m(E_\lambda^+) \leq \frac{512}{\pi}\left(q + \frac{q}{q-2}\right) m(E_\lambda^+) ,$$

which again is smaller than $\frac{1024}{\pi} q\, m(E_\lambda^+)$ provided that $q > 3$.

Using Hölder's inequality on (7.2) followed by the estimate above we get for $q > 3$ that

$$\lambda \cdot m(E_\lambda^+) \leq A^{1/q} \|H \chi_{E_\lambda^+}\|_p \leq A^{1/q} \cdot \left(\frac{1024}{\pi} q\right)^{1/p} [m(E_\lambda^+)]^{1/p}$$

which again gives

$$m(E_\lambda^+) \leq A \cdot \left(\frac{1}{\lambda}\right)^q \left(\frac{1024}{\pi} q\right)^{q/p} = \frac{A\pi}{1024 q} \cdot \left(\frac{1024}{\lambda\pi} q\right)^q .$$

If $\lambda > 3e \cdot \frac{1024}{\pi}$ we may put $q = \frac{\pi\lambda}{1024 e}$, so we get

$$m(E_\lambda^+) \leq \frac{Ae}{\lambda} \exp\left(-\frac{\pi\lambda}{1024 e}\right) .$$

Similarly we get

$$m(E_\lambda^-) \leq \frac{Ae}{\lambda} \exp\left(-\frac{\pi\lambda}{1024 e}\right) ,$$

and taking these inequalities together we have for large values of λ

$$m(\{x \mid |(H f)(x)| > \lambda\}) \leq \frac{2e A}{\pi} \cdot \exp\left(-\frac{\pi\lambda}{1024 e}\right)$$

proving theorem 7.1. □

Theorem 7.2. *There exist positive constants* c_1 *and* c_2 *such that, if* f *is any essentially bounded function, the support of which is contained in an interval of length* A *, then*

$$m(\{x \mid H^*f(x) > \lambda\}) \leqq c_1 A \frac{\|f\|_\infty}{\lambda} \exp\left(-c_2 \frac{\lambda}{\|f\|_\infty}\right) . \tag{7.3}$$

Proof. We remark that it suffices to prove (7.3) for large values of λ . We may assume that $f \geqq 0$ (write $f = f^+ - f^-$) and that $\|f\|_\infty = 1$. Let

$$E_\varepsilon^+ = \{x \mid \sup_{y \geqq \varepsilon} H_y f(x) > \lambda\} , \qquad E_\varepsilon^- = \{x \mid \sup_{y \geqq \varepsilon} (-H_y f(x)) > \lambda\} ,$$

where $\lambda > 0$ and $\varepsilon > 0$. We shall give an estimate for $m(E_\varepsilon^+)$, the estimate for $m(E_\varepsilon^-)$ being similar.

As in the proof of theorem 6.2 we can find disjoint intervals $I_1, I_2, \dots, I_n$, such that

$$m(E_\varepsilon^+) \leqq 4 \sum_{j=1}^n m(I_j) \tag{7.4}$$

and

$$\frac{1}{\pi}\int_{I_j^c} f(t) \frac{1}{c(I_j)-t} dt > \lambda , \qquad j = 1, 2, \dots , n . \tag{7.5}$$

Let $f_j(x) = f(x)\, \chi_{I_j}(x)$, $j = 1, \dots , n$, and consider the function

$$g_j(x) = H f(x) - H f_j(x) = \frac{1}{\pi}\int_{I_j^c} \frac{f(t)}{x-t} dt .$$

The function g_j is decreasing in I_j (because $f \geqq 0$) and $g_j(c(I_j)) > \lambda$ by (7.5), hence $g_j(x) > \lambda$ in the left half of I_j . Thus we infer that

$$\sum_{j=1}^n \frac{1}{2} m(I_j) \leqq m(\{x \mid H f(x) > \tfrac{\lambda}{2}\}) + \sum_{j=1}^n m(\{x \mid H f_j(x) < -\tfrac{\lambda}{2}\}) .$$

From theorem 7.1 we deduce that

$$\frac{1}{2}\sum_{j=1}^n m(I_j) \leqq c_1 A \cdot \frac{2}{\lambda} \exp\left(-c_2 \cdot \frac{\lambda}{2}\right) + \sum_{j=1}^n m(I_j) \cdot \left(\frac{2c_1}{\lambda} \exp\left(-c_2 \cdot \frac{\lambda}{2}\right)\right)$$

$$\leqq \frac{2c_1}{\lambda} A \exp\left(-\frac{c_2}{2}\lambda\right) + \frac{1}{4}\sum_{j=1}^n m(I_j) ,$$

provided λ is sufficiently large. Rearranging we then get

$$\sum_{j=1}^{n} m(I_j) \leqq \frac{8c_1}{\lambda} A \exp\left(-\frac{c_2}{2}\lambda\right) .$$

Combining this estimate with the inequality (7.4) we get

$$m(E_\varepsilon^+) \leqq \frac{32c_1}{\lambda} A \exp\left(-\frac{c_2}{2}\lambda\right) .$$

We have the same estimate for $m(E_\varepsilon^-)$ and thus

$$m(\{x \mid \sup_{y \geq \varepsilon} |H_y f(x)| > \lambda\}) \leqq \frac{64c_1}{\lambda} A \exp\left(-\frac{c_2}{2}\lambda\right)$$

The result then follows by letting $\varepsilon \to 0+$. □

CHAPTER III.

We shall now consider functions from $L^1(]-\pi,\pi])$. For technical reasons, however, we shall extend them by periodicity to the larger interval $]-4\pi,4\pi]$. Now a reasonable approach would be to consider smaller and smaller intervals, so we introduce the so-called dyadic intervals in § 8 depending on a rather stiff partition of the interval. Since we also need to take care of what is going on in the neighbouring intervals we introduce what we have called smoothing intervals. This partition of $]-4\pi,4\pi]$ into dyadic intervals is in some sense very convenient, but it means that we have to modify our different kinds of Hilbert transforms. This is also done in § 8 . In § 9 we generalize the concept of Fourier coefficients. The application of these also has a smoothing effect on the later results, especially in chapter IV. We prove some estimates for the generalized Fourier coefficients and for the constants derived from these. Finally, in § 10 , we shall review the operator M defined in § 4 . In order to give an estimate of M we define a new operator M^* , which is derived from some functions $S_n^*(x;f;\omega^*)$. [In the definition of $S_n^*(x;f;\omega^*)$ we use both the smoothing intervals introduced in § 8 and the Hilbert transform.] It follows that it is enough to consider the operator M^* , and we shall therefore give some preliminary estimates of $S_n^*(x;f;\omega^*)$. We cannot, however, prove all the estimates needed for $S_n^*(x;f;\omega^*)$. These are postponed to chapter IV.

§ 8. The dyadic intervals and the modified Hilbert transforms.

It follows from the formulation of theorem 4.2 which we still have to prove that we are only interested in periodic functions of period 2π , so in principle it would be enough to consider the spaces $L^p(]-\pi,\pi])$. When we use the Hilbert transform, however, we implicitly perform a convolution, which means that we also have to consider the periodic extension of f to $]-2\pi,2\pi]$. For some other technical reason we shall extend f to the larger interval $]-4\pi,4\pi]$.

For each $\nu \in N_o$ we divide $]-2\pi,2\pi]$ into $2 \cdot 2^\nu = 2^{\nu+1}$ half open intervals of equal length,

$$(8.1)\quad \omega_{j\nu} =]-2\pi + 2\pi(j-1)2^{-\nu}\ , -2\pi + 2\pi \cdot j \cdot 2^{-\nu}]\ , \qquad j = 1, 2, \dots , 2^{\nu+1}\ ,$$

where $m(\omega_{j\nu}) = 2\pi \cdot 2^{-\nu}$. These ω-intervals are in the following also called *dyadic intervals from level* $\nu \in N_o$.

It is natural to define $\omega_{1,-1} = \omega_{-1} =]-2\pi,2\pi]$ as the dyadic interval from level -1 . This interval ω_{-1} is only used once in chapter IV, so when nothing else is said we shall suppose that $\nu \in N_o$.

It is obvious that any $\omega_{j\nu}$ is the union of two neighbouring dyadic intervals from level $\nu+1$. The union of two neighbouring dyadic intervals from level $\nu+1$ need not, however, be a dyadic interval from level ν . We shall therefore introduce the so-called *smoothing intervals* from level ν (or intervals of ω^*-type) by

$$(8.2)\qquad \omega^*_{j\nu} = \omega_{j\nu} \cup \omega_{j+1,\nu} =]-2\pi + 2\pi(j-1)2^{-\nu}, -2\pi + 2\pi(j+1)2^{-\nu}]\ ,$$

where $\nu \in N_o$ and $j = 1, 2, \dots , 2^{\nu+1} - 1$. Thus

$$(8.3)\qquad m(\omega^*_{j\nu}) = 4\pi \cdot 2^{-\nu}\ .$$

For $\nu = -1$ we define $\omega^*_{-1} =]-4\pi, 4\pi]$. Note that $m(\omega^*_{-1}) = 8\pi$ in agreement with (8.3) for $\nu = -1$.

In the following we shall often use the short notation ω for any dyadic interval and ω^* for any smoothing interval.

By the construction it is obvious that we to any ω^* from level $\nu \in \{-1\} \cup N_0$ can find an ω from level $\nu + 1$, such that

$$(8.4) \qquad \omega \subset \omega^* \qquad \text{and} \qquad 4m(\omega) = m(\omega^*) .$$

Furthermore, ω can be chosen as one of the two intervals in the middle of ω^*. These simple facts are essential for the proof of theorem 4.2.

By the phrase "x belongs to the middle half of ω^*" we shall understand that x belongs to one of the two middle intervals of ω-type satisfying (8.4).

We shall now introduce the modified Hilbert transforms. These are necessary for two reasons. The first one is not very difficult to handle. In fact, we shall only consider a finite interval instead of the whole real axis, and it will only cause minor corrections in the results previously obtained. The second reason, however, is more essential for the proof, and it is also more difficult to obtain the crucial estimates. It is linked to our choice of method, because we shall only use the dyadic and the smoothing intervals introduced above, so the modified Hilbert transforms will be subordinated these rather stiff partitions of $]-2\pi, 2\pi]$.

Let $f \in L^1(]-\pi, \pi])$ and let f^0 be its periodic extension to $]-4\pi, 4\pi]$. Let ω^* be any smoothing interval from level $\nu \in N_0$. We define the *Hilbert transform* $H_{\omega^*} f$ *of* f *with respect to* ω^* by

$$(8.5) \qquad H_{\omega^*} f = H(f^0 \cdot \chi_{\omega^*}) ,$$

and the *maximal Hilbert transform* $H^*_{\omega^*} f$ *of* f *with respect to* ω^* by

$$(8.6) \qquad H^*_{\omega^*} f = H^*(f^0 \cdot \chi_{\omega^*}) .$$

It follows immediately from theorem 6.5 and theorem 6.7 that the operators H_{ω^*} and $H^*_{\omega^*}$ are of type p for all $p \in]1, +\infty[$, and even that the constants involved are independent of ω^*.

A closely related operator, which we shall denote by $\bar{H}$, is defined on $L^1(]-\pi,\pi])$ by

(8.7) $$\bar{H}f(x) = \sup_{\delta\in R_+} \left|\frac{1}{\pi}(pv)\int_{x-\delta}^{x+\delta}\frac{f^o(t)}{x-t}dt\right| .$$

<u>Lemma 8.1.</u> *The operator* $\bar{H}$ *is of type* p *for all* $p\in]1,+\infty[$.

<u>Proof.</u> It follows from the identity

$$\frac{1}{\pi}(pv)\int_{x-\delta}^{x+\delta}\frac{f^o(t)}{x-t}dt = \frac{1}{\pi}(pv)\int_{-\infty}^{+\infty}\frac{f^o(t)}{x-t}dt - \frac{1}{\pi}\int_{\{|x-t|\geq\delta\}}\frac{f^o(t)}{x-t}dt$$

that

(8.8) $$\bar{H}f(x) \leq |Hf^o(x)| + H^*f^o(x) .$$

As $\|Hf^o\|_p \leq C_p\|f^o\|_p$ and $\|H^*f^o\|_p \leq C_p^*\|f^o\|_p$, say, and $\|f^o\|_p = 4^{1/p}\|f\|_p$, we get from (8.8) for $p\in]1,+\infty[$,

$$\|\bar{H}f\|_p \leq \|Hf^o\|_p + \|H^*f^o\|_p \leq 4^{1/p}(C_p+C_p^*)\|f\|_p . \qquad \square$$

Finally, we shall introduce the modified maximal Hilbert transform $\hat{H}$ subordinated the dyadic intervals introduced above.

Consider a given $\omega_{j\nu}^*$ and let x be an interior point of $\omega_{j\nu}^*$, $x\in\operatorname{int}\omega_{j\nu}^*$. Then there exists a uniquely determined sequence $(\sigma_x^r)_{r=\mu,\mu+1,\dots}$ of intervals, such that each σ_x^r is a smoothing interval from level r , $\sigma_x^r\subset\omega^*$, and such that x belongs to the middle half of each σ_x^r . Note that μ depends on x and that $\mu\geq\nu$.

The *modified maximal Hilbert transform* $\hat{H}_{j\nu}$ *with respect to* $\omega_{j\nu}^*$ is defined by

(8.9) $$\hat{H}_{j\nu}f(x) = \sup_{r\geq\mu}\left|\frac{1}{\pi}(pv)\int_{\sigma_x^r}\frac{f^o(t)}{x-t}dt\right| , \qquad x\in\operatorname{int}\omega_{j\nu}^* ,$$

where $f\in L^1(]-\pi,\pi])$ and where f^o as usual denotes the periodic extension of f to $]-4\pi, 4\pi]$. We shall sometimes for short write (8.9) in the form

$$\hat{H} f(x) = \sup_{\sigma_x} \left| \frac{1}{\pi} (pv) \int_{\sigma_x} \frac{f^o(t)}{x-t} dt \right| .$$

We shall prove that $\hat{H}$ also is of type p for every $p \in]1,+\infty[$. As σ_x^r normally is a skew interval with respect to x , we shall first prove the following lemma.

Lemma 8.2. *Let $f \in L^1(]-\pi, \pi])$ and let f^o be the periodic extension of f to $]-4\pi, 4\pi]$. By $\mathfrak{J}$ we shall denote the family of intervals $\Delta =]-a, b[$, $-\pi \leq -a < 0 < b \leq \pi$, satisfying the condition $\frac{1}{3} \leq \frac{a}{b} \leq 3$, and by $\mathfrak{J}^*$ we shall denote the family of symmetric intervals $I =]-\delta, \delta[$ contained in $]-\pi, \pi[$. Then*

$$(8.10) \quad \sup_{\Delta \in \mathfrak{J}} \left| \frac{1}{\pi} (pv) \int_\Delta \frac{f^o(x+t)}{t} dt \right| \leq \sup_{I \in \mathfrak{J}^*} \left| \frac{1}{\pi} (pv) \int_I \frac{f^o(x+t)}{t} dt \right| + 3\Theta f^o(x) .$$

Proof. Let $\Delta =]-a, b[\in \mathfrak{J}$, where $0 < a < b$, so $1 < \frac{b}{a} \leq 3$. Then

$$(8.11) \quad \left| \frac{1}{\pi} (pv) \int_\Delta \frac{f^o(x+t)}{t} dt \right| \leq \left| \frac{1}{\pi} (pv) \int_{-a}^{a} \frac{f^o(x+t)}{t} dt \right| + \left| \frac{1}{\pi} \int_a^b \frac{f^o(x+t)}{t} dt \right| ,$$

where $]-a, a[\in \mathfrak{J}^*$. Let $F_x(t) = \int_0^t f^o(x+y)dy$. By a partial integration we get

$$\frac{1}{\pi} \int_a^b \frac{f^o(x+t)}{t} dt = \frac{1}{\pi} \left[\frac{F_x(t)}{t} \right]_a^b + \frac{1}{\pi} \int_a^b \frac{F_x(t)}{t^2} dt .$$

Using that $|F_x(t)| \leq 2|t| \Theta f^o(x)$ [cf. lemma 2.4] we deduce the estimate

$$\left| \frac{1}{\pi} \int_a^b \frac{f^o(x+t)}{t} dt \right| \leq \frac{1}{\pi} \left\{ \frac{1}{b} |F_x(b)| + \frac{1}{a} |F_x(a)| \right\} + \frac{1}{\pi} \int_a^b \frac{2t \, \Theta f^o(x)}{t^2} dt$$

$$\leq \frac{4}{\pi} \Theta f^o(x) + \frac{2}{\pi} \Theta f^o(x) \int_a^b \frac{dt}{t} = \frac{2}{\pi} \Theta f^o(x) \cdot \left\{ 2 + \log \frac{b}{a} \right\} \leq \frac{2}{\pi} (2 + \log 3) \Theta f^o(x) \leq 3\Theta f^o(x) .$$

When this inequality is substituted into (8.11) we easily get (8.10). The case $1 \leq \frac{a}{b} \leq 3$ is treated similarly. □

Theorem 8.3. *The modified maximal Hilbert transform $\hat{H}$ with respect to ω^* is of type p for all $p \in]1,+\infty[$.*

Proof. Let $x \in \operatorname{int} \omega^*$ and let σ_x be any of the intervals associated with x and ω^* in (8.9). Then $\sigma_x - x = \Delta$ is an interval from the family $\mathfrak{J}$ introduced in lemma 8.2, so using lemma 8.2 we deduce that

$$\hat{H} f(x) \leqq \bar{H} f^{o}(x) + 3\,\Theta f^{o}(x) . \tag{8.12}$$

As both $\bar{H}$ (cf. lemma 8.1) and Θ are of type p for all $p \in]1,+\infty[$ we easily deduce that $\hat{H}$ also is of type p , following the same line as the proof of lemma 8.1. □

It follows immediately from (8.8) and (8.12) that

$$\hat{H}_{\omega^*} f(x) \leqq |H f^{o}(x)| + H^* f^{o}(x) + 3\,\Theta f^{o}(x)$$

for any $\omega^* \neq \omega^*_{-1}$, so $\{x \in \omega^* \mid \hat{H} f(x) > \lambda\}$ is contained in

$$\{x \in \omega^* \mid |H f^{o}(x)| > \tfrac{\lambda}{3}\} \cup \{x \in \omega^* \mid H^* f^{o}(x) > \tfrac{\lambda}{3}\} \cup \{x \in \omega^* \mid \Theta f^{o}(x) > \tfrac{\lambda}{9}\} .$$

With trivial modifications this is also true for $\omega^* = \omega^*_{-1}$. Thus, if $f \in L^{\infty}$ it follows immediately from theorem 2.6, theorem 7.1 and theorem 7.2 that we also have an exponential estimate for $\hat{H}_{\omega^*}$. In fact,

Theorem 8.4. *There exist positive constants c_1 and c_2 such that, if f is any essentially bounded function and $\hat{H}$ is the modified maximal Hilbert transform with respect to ω^* , then*

$$m(\{x \in \omega^* \mid \hat{H} f(x) > \lambda\}) \leqq c_1 \cdot m(\omega^*) \cdot \frac{\|f\|_\infty}{\lambda} \cdot \exp\left(-c_2 \frac{\lambda}{\|f\|_\infty}\right) . \tag{8.13}$$

We note that $m(\{x \in \omega^* \mid \hat{H} f(x) > \lambda\}) \leqq m(\omega^*)$, and as the function $\varphi(t) = \frac{1}{t} \exp(-c_2 t)$, $t \in R_+$, is decreasing and $\varphi(t) \leqq \exp(-c_2 t)$ for $t \geqq 1$, we can omit the factor $\frac{1}{t} = \frac{\|f\|_\infty}{\lambda}$ in front of the exponential function on the right hand side of (8.13), provided that $c_1 \geqq \exp c_2$. Thus we also have the following corollary.

Corollary 8.5. *There exist positive constants c_1 and c_2 such that, if f is any essentially bounded function and $\hat{H}$ is the modified maximal Hilbert transform with respect to ω^*, then*

$$m(\{x \in \omega^* \mid \hat{H} f(x) > \lambda\} \leqq c_1 m(\omega^*) \exp\left(-c_2 \cdot \frac{\lambda}{\|f\|_\infty}\right) .$$

The constants c_1 and c_2 do not depend on the choice of ω^*.

§ 9. Generalized Fourier coefficients.

In the following ω will always denote a dyadic interval as introduced in § 8, and $\omega_{j\nu}$ will denote one of the dyadic intervals of length $2\pi \cdot 2^{-\nu}$ contained in $]-2\pi, 2\pi]$. Similarly, ω^* will denote a smoothing interval (including $\omega^*_{-1} =]-4\pi, 4\pi])$, and $\omega^*_{j\nu}$ will denote one of the smoothing intervals of length $4\pi \cdot 2^{-\nu}$.

For each $n \in \mathbb{N}_0$ and each dyadic interval $\omega_{j\nu}$ we define

$$(9.1) \qquad \psi(n; \omega_{j\nu}) = \left[\frac{n}{2\pi} \cdot m(\omega_{j\nu})\right] = [n \cdot 2^{-\nu}] ,$$

where $[n \cdot 2^{-\nu}]$ denotes the greatest non-negative integer less than or equal to $n \cdot 2^{-\nu}$.

If $\omega^*_{j\nu}$ is any smoothing interval from level $\nu \in \mathbb{N}_0$, let $\omega_{k,\nu+1}$ be any of the four dyadic intervals satisfying (8.4). We define

$$(9.2) \qquad \psi^*(n; \omega^*_{j\nu}) = \psi(n; \omega_{k,\nu+1}) = [n \cdot 2^{-\nu-1}] = [\tfrac{n}{8\pi} m(\omega^*_{j\nu})] .$$

For $\omega^* = \omega^*_{-1}$ we define $\psi^*(n; \omega^*_{-1}) = n$, which is consistent with (9.2).

Remark 9.1. Here we have changed the usual notation, which is $n[\omega_{j\nu}]$ for $\psi(n; \omega_{j\nu})$ and $n^*[\omega^*_{j\nu}]$ for $\psi^*(n; \omega^*_{j\nu})$. This has been done in order to avoid confusion. For the same reason we have avoided the use of $|\cdot|$ for the Lebesgue measure, since it may be difficult to distinguish between $n[\omega]$ and $n|\omega|$. In our notation they are written $\psi(n; \omega)$ and $n \cdot m(\omega)$. In chapter IV they will both occur in formulæ, which are very much alike.

For an arbitrary function $f \in L^1(]-\pi,\pi])$ we let (as before) f^o denote the periodic extension of f to $]-4\pi, 4\pi]$ with period 2π.

We define, for $\alpha \in R$, $\omega = \omega_{j\nu}$ and $f \in L^1(]-\pi,\pi])$, the *α'th generalized Fourier coefficient* $c_\alpha(\omega; f)$ by

$$(9.3) \quad c_\alpha(\omega;f) = \frac{1}{m(\omega)} \int_\omega f^o(x) \exp(-i\, 2^\nu \alpha x)dx = \frac{1}{m(\omega)} \int_\omega f^o(x) \exp\left(-i\frac{2\pi\alpha x}{m(\omega)}\right)dx .$$

We note that if $\alpha = n \in Z$ we have the ordinary Fourier coefficients of f (with respect to the interval ω), but in the following it will be necessary also to deal with the generalized coefficient, especially with the case where $\alpha = \frac{\mu}{3}$, $\mu \in Z$.

We first remark that we of course have the estimate

$$(9.4) \qquad |c_\alpha(\omega;f)| \leq \frac{1}{m(\omega)} \int_\omega |f^o(x)|dx \leq \frac{2^\nu}{2\pi} \|f\|_1 .$$

Next we notice that

$$(9.5) \qquad \sum_{\mu=-\infty}^{+\infty} \frac{1}{1+\mu^2} < 10 ,$$

which ensures that for $n \in Z$ we can define

$$(9.6) \qquad C_n(\omega;f) = \frac{1}{10} \sum_{\mu=-\infty}^{+\infty} |c_{n+\frac{\mu}{3}}(\omega;f)| \cdot \frac{1}{1+\mu^2} ,$$

and we have from the inequalities above that

$$(9.7) \qquad 0 \leq C_n(\omega;f) \leq \sup_{\mu\in Z} |c_{n+\frac{\mu}{3}}(\omega;f)| \leq \frac{1}{m(\omega)} \int_\omega |f^o(x)|dx .$$

We also have to introduce the constants $C_n^*(\omega^*;f)$. This is done in the following way. First we consider ω_{-1}^*. We put

$$(9.8) \qquad C_n^*(\omega_{-1}^*;f) = C_n(\omega_{1,0};f) = C_n(\omega_{2,0};f) .$$

(The latter equation follows from the fact that f^o is a periodic extension of f and that $\omega_{1,0} =]-2\pi, 0]$ and $\omega_{2,0} =]0, 2\pi]$.)

For all other intervals of ω^*-type we put

(9.9) $$C_n^*(\omega^*;f) = \max\{C_n(\omega;f) \mid \omega \subset \omega^*,\ 4m(\omega) = m(\omega^*)\}\ ,$$

i.e. $C_n^*(\omega^*;f)$ is the largest of four well-defined numbers, as there are exactly four intervals of ω-type satisfying (8.4).

We note that the functions $\psi(n;\omega)$ and $\psi^*(n;\omega^*)$, $n \in N_0$, introduced in (9.1) and (9.2) , are defined in such a way that

(9.10) $$C^*_{\psi^*(n;\omega^*)}(\omega^*;f) = \max\{C_{\psi(n;\omega)}(\omega;f) \mid \omega \subseteqq \omega^*,\ 4m(\omega) = m(\omega^*)\}\ .$$

If the function f is equal to 0 almost everywhere in ω , then of course all coefficients $c_\alpha(\omega;f)$ and all numbers $C_n(\omega;f)$ are equal to 0 . Conversely, if there exists an $n \in Z$ such that $C_n(\omega;f) = 0$ we infer from (9.6) that $c_{n+\frac{\mu}{3}}(\omega;f) = 0$ for all $\mu \in Z$, i.e. $c_{n+p+(\mu-3p)/3}(\omega;f) = 0$ for all $\mu - 3p \in Z$ and every fixed $p \in Z$. This shows that $C_n(\omega;f) = 0$ for *all* $n \in Z$ provided that just one of the $C_n(\omega;f)$ equals 0 .

Furthermore we conclude from the condition

$$c_{n+\frac{\mu}{3}}(\omega;f) = \frac{1}{m(\omega)}\int_\omega f^o(x)\exp\left(-i\frac{2\pi}{m(\omega)}\{n+\frac{\mu}{3}\}x\right)dx = 0$$

for all $\mu \in Z$ (actually it suffices with $\mu \in 3Z$) that $f^o(x) = 0$ almost everywhere in ω .

We have thus proved the equivalence of the following three conditions:

(9.11) $f^o(x) = 0$ *almost everywhere in* ω ;

(9.12) *there exists an* $n \in Z$ *such that* $C_n(\omega;f) = 0$;

(9.13) $C_n(\omega;f) = 0$ *for all* $n \in Z$.

We shall now prove some results concerning the coefficients $C_n(\omega;f)$, which will be needed later on. In order to shorten the notation we shall sometimes in the proofs write c_α or $c_\alpha(\omega)$ instead of $c_\alpha(\omega;f)$, and C_n or $C_n(\omega)$ instead of $C_n(\omega;f)$ where no misunderstanding is possible.

First we shall prove a general lemma. In what follows we shall need some positive constants. These will be denoted by $c_1, c_2, \ldots$ and should not be confused with the shortened notation c_α or c_n for the generalized Fourier coefficients.

Lemma 9.2. *To any* $\varphi \in C^2([0,2\pi])$ *there exist polynomials* $p_1(t)$ *and* $p_2(t)$ *, such that the function* $\Phi(t)$ *, defined by*

$$\Phi(t) = \begin{cases} p_1(t) & \text{for } x \in [-2\pi,0] \\ \varphi(t) & \text{for } x \in [0,2\pi] \\ p_2(t) & \text{for } x \in [2\pi,4\pi] \end{cases}$$

satisfies the following conditions

$$\Phi \in C^2([-2\pi,4\pi]) , \qquad \Phi^{(k)}(-2\pi) = \Phi^{(k)}(4\pi) = 0 \quad \text{for } k=0,1,2 ,$$

and for some constant $c_1 > 1$

$$\max_{[-2\pi,4\pi]} |\Phi(t)| + \max_{[-2\pi,4\pi]} |\Phi''(t)| \leq c_1\{\|\varphi\|_\infty + \|\varphi''\|_\infty\} , \tag{9.14}$$

where $\|\psi\|_\infty = \max_{[0,2\pi]} |\psi(t)|$ *for any function* $\psi \in C^0([0,2\pi])$.

Proof. We shall prove (9.14) by proving it in each of the intervals $[-2\pi,0]$, $[0,2\pi]$ and $[2\pi,4\pi]$. It is trivial in $[0,2\pi]$. It is easy to construct a polynomial $p_1(t)$ of degree 5 in $[-2\pi,0]$, such that $p_1^{(k)}(-2\pi) = 0$ for $k=0,1,2,$ and $p_1^{(k)}(0) = \varphi^{(k)}(0)$, $k=0,1,2$, so Φ is of class C^2 also for $t=0$. This polynomial $p_1(t)$ only depends on $\varphi''(0)$, $\varphi'(0)$ and $\varphi(0)$, which may be arbitrary numbers, so there exist constants, such that

$$|p_1(t)| \leq c_0'|\varphi(0)| + c_1'|\varphi'(0)| + c_2'|\varphi''(0)| , \tag{9.15}$$

$$|p_1''(t)| \leq c_0''|\varphi(0)| + c_1''|\varphi'(0)| + c_2''|\varphi''(0)| . \tag{9.16}$$

In order to estimate $|\varphi'(0)|$ we use the following Poincaré-like method. From the equation

$$\int_0^{2\pi} (t-2\pi)\,\varphi''(t)dt = [(t-2\pi)\,\varphi'(t)]_0^{2\pi} - \int_0^{2\pi} \varphi'(t)dt = 2\pi\,\varphi'(0) - \{\varphi(2\pi) - \varphi(0)\}$$

we get after a rearrangement

$$|\varphi'(0)| \leq \frac{1}{2\pi}\left\{2\|\varphi\|_\infty + \|\varphi''\|_\infty \int_0^{2\pi} |t-2\pi|dt\right\} = \frac{1}{\pi}\left\{\|\varphi\|_\infty + \pi^2\|\varphi''\|_\infty\right\}$$

When we substitute this inequality and $|\varphi(0)| \leq \|\varphi\|_\infty$ and $|\varphi''(0)| \leq \|\varphi''\|_\infty$ into (9.15) and (9.16) we easily derive (9.14) in the interval $[-2\pi,0]$ for a suitable $c_1 > 1$.

As $p_2(t)$ in $[2\pi,4\pi]$ is defined in a similar manner, the lemma is proved. □

Remark 9.3. If we choose

$$p_1(t) = \frac{1}{16\pi^3}(t+2\pi)^3\,t^2\,\varphi''(0) + \frac{1}{16\pi^4}(t+2\pi)^3\,t\cdot(2\pi-3t)\,\varphi'(0)$$

$$+ \frac{1}{16\pi^5}(t+2\pi)^3\,(3t^2-3\pi\,t+2\pi^2)\,\varphi(0)$$

in $[-2\pi,0]$ and let $p_2(t)$ be similarly defined in $[2\pi,4\pi]$, it is possible to prove that Φ satisfies the conditions of the theorem, and that

$$\max_{[-2\pi,4\pi]} |\Phi(t)| + \max_{[-2\pi,4\pi]} |\Phi''(t)| \leq 2\|\varphi\|_\infty + 8\|\varphi''\|_\infty \ .$$

Lemma 9.4. *Let* $\varphi \in C^2(\bar{\omega})$, $\omega = \omega_{j\nu}$. *Then*

$$\varphi(t) = \sum_{\mu\in Z} \gamma_\mu \exp\left(i\,2^\nu\cdot\frac{\mu}{3}t\right) , \qquad t \in \omega ,$$

where for some constant c_2 ,

$$(1+\mu^2)\,|\gamma_\mu| \leq c_2 \cdot \{\max_\omega|\varphi| + 2^{-2\nu}\max_\omega|\varphi''|\} \ .$$

Proof. We may assume that $\bar{\omega} = [0,2\pi]$, using if necessary the change of variable $t = 2^{-\nu}\tau + \alpha$. Let Φ be the function introduced in lemma 9.2. If we expand Φ in a Fourier series over $[-2\pi, 4\pi]$ we get

$$\Phi(t) = \sum_{\mu\in Z} \gamma_\mu \exp\left(i\frac{\mu}{3}t\right) , \quad t\in[-2\pi, 4\pi] ,$$

and especially,

$$\varphi(t) = \sum_{\mu\in Z} \gamma_\mu \exp\left(i\frac{\mu}{3}t\right) , \quad t\in[0,2\pi] .$$

Furthermore, we have

$$\Phi''(t) = \sum_{\mu\in Z} \left(-\frac{\mu^2}{9}\right)\gamma_\mu \exp\left(i\frac{\mu}{3}t\right), \quad t\in[-2\pi, 4\pi] .$$

This gives that $|\gamma_\mu| \leq \max_{[-2\pi,4\pi]} |\Phi|$ and $\frac{\mu^2}{9}|\gamma_\mu| \leq \max_{[-2\pi,4\pi]} |\Phi''|$ for $\mu\in Z$, and so according to lemma 9.2,

$$(1+\mu^2)|\gamma_\mu| \leq 9\{\max_{[-2\pi,4\pi]} |\Phi| + \max_{[-2\pi,4\pi]} |\Phi''|\} \leq 9c_1\{\max_{[0,2\pi]} |\varphi| + \max_{[0,2\pi]} |\varphi''|\} ,$$

so we may choose $c_2 = 9c_1$. The factor $2^{-2\nu}$ in the lemma is due to the change of variable and the fact that we differentiate twice. □

Lemma 9.5. *There exists a constant* $c_3 > 0$ *such that we for* $n\in N_0$ *and* $\omega = \omega_{j\nu}$ *have*

$$|c_{n\cdot 2^{-\nu}}(\omega;f)| \leq c_3 \cdot C_{\psi(n;\omega)}(\omega;f) .$$

Proof. Let $\beta = n - \psi(n;\omega)\cdot 2^\nu = n - 2^\nu[n\cdot 2^{-\nu}]$. Then we have $0 \leq \beta < 2^\nu$. Using lemma 9.4 on the function $\varphi(t) = e^{-i\beta t}$ and $\omega = \omega_{j\nu}$ we get for $t\in\omega_{j\nu}$,

$$\varphi(t) = e^{-i\beta t} = \sum_{\mu\in Z} \gamma_\mu \exp\left(i\, 2^\nu\cdot\frac{\mu}{3}t\right) = \sum_{\mu\in Z} \bar{\gamma}_\mu \exp\left(-i\cdot 2^\nu\cdot\frac{\mu}{3}t\right)$$

with

$$(1+\mu^2)|\gamma_\mu| \leq c_2\cdot\{\max_\omega|\varphi| + 2^{-2\nu}\max_\omega|\varphi''|\} \leq c_2\cdot\{1+2^{-2\nu}\beta^2\} \leq c_2\{1+2^{-2\nu}\cdot 2^{2\nu}\} = 2c_2 .$$

Then we have

$$c_{n\cdot 2^{-\nu}}(\omega;f) = \frac{1}{m(\omega)}\int_\omega e^{-int} f^o(t)dt = \frac{1}{m(\omega)}\int_\omega e^{-i\beta t}\exp[-i\psi(n;\omega)2^\nu t]\cdot f^o(t)dt$$

$$= \sum_{\mu\in Z}\tilde{\gamma}_\mu\, c_{\psi(n;\omega)+\frac{\mu}{3}}(\omega;f) ,$$

and hence

$$|c_{n\cdot 2^{-\nu}}(\omega;f)| \leqq \sum_{\mu\in Z}|\gamma_\mu|\cdot|c_{\psi(n;\omega)+\frac{\mu}{3}}(\omega;f)| \leqq 20c_2\cdot\frac{1}{10}\sum_{\mu\in Z}|c_{\psi(n;\omega)+\frac{\mu}{3}}(\omega;f)|\cdot\frac{1}{1+\mu^2}$$

$$= 20c_2\cdot C_{\psi(n;\omega)}(\omega;f)$$

proving the lemma. □

Lemma 9.6. *Let* $n\in Z$ *and* $f\in L^2(\omega)$ *be given where* $\omega=\omega_{j\nu}$. *Let* $A,B\in R_+$ *and* $M\geqq 2$ *be constants such that*

$$\int_\omega |f(t)|^2dt \leqq A^2\, m(\omega)$$

and

$$|c_m(\omega;f)| \leqq B \qquad \text{for} \quad |n-m| \leqq M .$$

Then we have

$$C_n(\omega;f) \leqq 9\cdot\left\{\frac{A}{\sqrt{M}} + B\log M\right\} .$$

Proof. We have

$$C_n(\omega;f) = \frac{1}{10}\sum_{\mu\in Z}|c_{n+\frac{\mu}{3}}(\omega;f)|\cdot\frac{1}{1+\mu^2}$$

and

$$c_{n+\frac{\mu}{3}}(\omega;f) = \frac{1}{m(\omega)}\int_\omega f(x)\cdot\exp\left(-i\,2^\nu(n+\tfrac{\mu}{3})x\right)dx = \frac{1}{m(\omega)}\int_\omega f(x)\cdot\exp(-i\,\alpha_\mu x)dx ,$$

where $\alpha_\mu = 2^\nu(n+\frac{\mu}{3})$.

By a Fourier expansion we get

$$\exp(i\,\alpha_\mu x) = \sum_{k\in Z}\alpha_{\mu,k}\cdot\exp(i\,2^\nu kx) ,$$

where

$$\alpha_{\mu,k} = \frac{1}{m(\omega)} \int_{\omega} \exp\left(i(\alpha_\mu - 2^\nu k)x\right) dx = \frac{1}{m(\omega)} \int_{\omega} \exp\left(- i\, 2^\nu \{k - (n+\tfrac{\mu}{3})\}x\right) dx \;.$$

By an integration we get that $\alpha_{\mu,k} = 0$ if $k - (n + \frac{\mu}{3}) \in Z\setminus\{0\}$, that $\alpha_{\mu,k} = 1$ if $k - (n + \frac{\mu}{3}) = 0$ and that

$$|\alpha_{\mu,k}| \leqq \frac{2}{2\pi} \cdot |k - (n + \tfrac{\mu}{3})|^{-1} \qquad \text{if} \quad k - (n + \tfrac{\mu}{3}) \notin Z \;.$$

Hence we have

$$|\alpha_{\mu,k}| \leqq \frac{4}{\pi} \cdot \frac{1}{1 + |k - (n + \frac{\mu}{3})|} \,, \qquad \mu\,, k \in Z \;.$$

As $f(x) = \sum_{k\in Z} c_k(\omega;f) \exp[i\, 2^\nu kx]$ (in L^2-sense) we get

$$\begin{aligned} c_{n+\frac{\mu}{3}}(\omega;f) &= \frac{1}{m(\omega)} \int_{\omega} \sum_{k\in Z} c_k(\omega;f) \exp(i\, 2^\nu kx) \cdot \overline{\exp(i\, \alpha_\mu x)}\, dx \\ &= \sum_{k\in Z} \frac{1}{m(\omega)} \int_{\omega} c_k(\omega;f) \exp(i\, 2^\nu kx) \sum_{\ell\in Z} \bar{\alpha}_{\mu,\ell} \exp(-i\, 2^\nu \ell x)\, dx \\ &= \sum_{k\in Z} c_k(\omega;f)\, \bar{\alpha}_{\mu,k} \;, \end{aligned}$$

and we have the estimate

$$|c_{n+\frac{\mu}{3}}(\omega;f)| \leqq \sum_{k\in Z} |c_k(\omega;f)| \cdot |\alpha_{\mu,k}| \leqq \frac{4}{\pi} \sum_{k\in Z} |c_k(\omega;f)| \cdot \frac{1}{1 + |k - (n + \frac{\mu}{3})|} \;.$$

We now consider the two cases: 1) $|\frac{\mu}{3}| \leqq \frac{1}{2}M$ and 2) $|\frac{\mu}{3}| > \frac{1}{2}M$. In the first case we get as $M \geqq 2$

$$\begin{aligned} \sum_{|k-n|\leqq M} |c_k(\omega;f)| \cdot \frac{1}{1 + |k - n - \frac{\mu}{3}|} &\leqq B \sum_{|k-n|\leqq M} \frac{1}{1 + |k - n - \frac{\mu}{3}|} \leqq B \sum_{1\leqq|\ell|\leqq 2M} \frac{1}{|\ell|} \\ &\leqq 2B \cdot (1 + \log 2M) \leqq 8B \cdot \log M \;, \end{aligned}$$

and

$$\sum_{|k-n|>M} |c_k(\omega;f)| \cdot \frac{1}{1+|k-n-\frac{\mu}{3}|} \leqq \left\{\sum_k |c_k(\omega;f)|^2\right\}^{\frac{1}{2}} \cdot \left\{\sum_{|k-n|>M} \frac{1}{(1+|k-n-\frac{\mu}{3}|)^2}\right\}^{\frac{1}{2}}$$

$$\leqq A \cdot \left\{2 \sum_{\ell > \frac{M}{2}} \frac{1}{\ell^2}\right\}^{\frac{1}{2}} \leqq \sqrt{8} \cdot A \cdot \frac{1}{\sqrt{M}} \quad .$$

Hence we get

$$|c_{n+\frac{\mu}{3}}(\omega;f)| \leqq 8 \cdot \left\{\frac{A}{\sqrt{M}} + B \cdot \log M\right\} = q \qquad \text{for} \quad |\tfrac{\mu}{3}| \leqq \tfrac{M}{2} .$$

In the second case $\left(|\frac{\mu}{3}| > \frac{M}{2}\right)$ we use the estimate $|c_{n+\frac{\mu}{3}}(\omega;f)| \leqq A$. Then we get

$$C_n(\omega;f) \leqq \frac{1}{10} \sum_{|\frac{\mu}{3}| \leqq \frac{M}{2}} |c_{n+\frac{\mu}{3}}| \cdot \frac{1}{1+\mu^2} + \frac{1}{10} \sum_{|\frac{\mu}{3}| > \frac{M}{2}} |c_{n+\frac{\mu}{3}}| \cdot \frac{1}{1+\mu^2}$$

$$\leqq q \cdot \frac{1}{10} \sum_{\mu\in Z} \frac{1}{1+\mu^2} + A \cdot \frac{1}{10} \cdot \sum_{|\frac{\mu}{3}| > \frac{M}{2}} \frac{1}{1+\mu^2} \leqq q + \frac{1}{5} \cdot \frac{A}{M} \leqq q + \frac{1}{5} A \cdot \frac{1}{\sqrt{M}} \quad .$$

Substituting the value of q we get the estimate

$$C_n(\omega;f) \leqq 9\left\{\frac{A}{\sqrt{M}} + B \log M\right\} ,$$

and the lemma is proved. □

Lemma 9.7. *If $f(x) = e^{i\lambda x}$ and $\omega = \omega_{j\nu}$ then*

$$|2^{-\nu}\lambda - n| \cdot C_n(\omega;f) \leqq 1 .$$

Proof. As $|f(x)| = 1$ we have $|c_{n+\frac{\mu}{3}}(\omega;f)| \leqq 1$ and hence also $C_n(\omega;f) \leqq 1$ for all $n \in Z$. It is therefore enough to consider the case where $|2^{-\nu}\lambda - n| \geqq 1$.

We again consider two cases:

$$1) \quad |\tfrac{\mu}{3}| \leqq \tfrac{1}{2}|2^{-\nu}\lambda - n| \qquad \text{and} \qquad 2) \quad |\tfrac{\mu}{3}| > \tfrac{1}{2}|2^{-\nu}\lambda - n| .$$

In the first case we have

$$|c_{n+\frac{\mu}{3}}(\omega;f)| = \left|\frac{1}{m(\omega)}\int_\omega \exp(i\lambda x)\exp\left(-i\,2^\nu(n+\tfrac{\mu}{3})x\right)dx\right|$$

$$= \left|\frac{1}{m(\omega)}\int_\omega \exp\left(i\,2^\nu\{2^{-\nu}\lambda - n - \tfrac{\mu}{3}\}x\right)dx\right| ,$$

which is smaller than $\frac{2}{2\pi}\cdot\frac{1}{|2^{-\nu}\lambda - n - \frac{\mu}{3}|}$, so we get

$$|2^{-\nu}\lambda - n|\cdot|c_{n+\frac{\mu}{3}}| \leqq |2^{-\nu}\lambda - n|\cdot\frac{2}{2\pi}\cdot\frac{1}{|2^{-\nu}\lambda - n - \frac{\mu}{3}|} \leqq \frac{2}{\pi} .$$

In the second case, where $|\frac{\mu}{3}| > \frac{1}{2}|2^{-\nu}\lambda - n|$, we use that $|c_{n+\frac{\mu}{3}}(\omega;f)| \leqq 1$, and we get

$$|2^{-\nu}\lambda - n|\,C_n(\omega;f) \leqq |2^{-\nu}\lambda - n|\cdot\frac{1}{10}\cdot\sum_{|\frac{\mu}{3}|\leqq\frac{1}{2}|2^{-\nu}\lambda-n|}|c_{n+\frac{\mu}{3}}|\cdot\frac{1}{1+\mu^2}$$

$$+\ |2^{-\nu}\lambda - n|\cdot\frac{1}{10}\cdot\sum_{|\frac{\mu}{3}|>\frac{1}{2}|2^{-\nu}\lambda-n|}|c_{n+\frac{\mu}{3}}|\cdot\frac{1}{1+\mu^2}$$

$$\leqq \frac{2}{\pi}\cdot\frac{1}{10}\sum_{\mu\in Z}\frac{1}{1+\mu^2} + |2^{-\nu}\lambda - n|\cdot\frac{1}{10}\cdot\sum_{|\frac{\mu}{3}|>\frac{1}{2}|2^{-\nu}\lambda-n|}\frac{1}{1+\mu^2} \leqq \frac{2}{\pi}+\frac{1}{5} < 1 ,$$

proving the lemma. □

§ 10. The functions $S_n^*(x;f;\omega^*)$ and the operator M^* .

Let $p \in]1,+\infty[$ and let $\mathfrak{M}$ denote the class of measurable functions with values in $[0,+\infty]$. In § 4 we defined an operator $M : L^p([-\pi,\pi]) \to \mathfrak{M}$ by

$$Mf(x) = \sup_{n\in N_0}|S_n(x;f)| ,$$

where $S_n(x;f)$ is the partial sum

$$S_n(x;f) = \sum_{k=-n}^{n} c_k e^{ikx} , \qquad x \in [-\pi,\pi] , \qquad n \in N_0 ,$$

of the Fourier series for f . We note that

(10.1) $$S_n(x;f) - S_{n-1}(x;f) = c_n e^{inx} + c_{-n} e^{-inx}, \quad n \in N .$$

We still have to prove theorem 4.2, i.e. that M is of type p for all $p \in]1,+\infty[$. Instead of giving a direct proof we shall in this section introduce another operator $M^* : L^p([-\pi,\pi]) \to \mathfrak{M}$ and prove that

(10.2) $$\|Mf\|_p \leq 5\|f\|_p + \|M^*f\|_p ,$$

where $\|\cdot\|_p$ always denotes the p-norm of $L^p([-\pi,\pi])$. The operator M^* is dealt with in chapter IV. We may note that we do not prove that M^* itself is of type p , but only of restricted type p , $p \in]1,+\infty[$.

If $f \in L^1(]-\pi,\pi])$ is any real valued function from L^1 we always let $f^o \in L^1(]-4\pi,4\pi])$ denote the periodic extension of f to $\omega^*_{-1} =]-4\pi,4\pi]$.

The definition of M^* depends heavily on the Hilbert transform. Let ω^* be any interval of smoothing type introduced in § 8 , and let $n \in Z$. We define the functions $S^*_n(x;f;\omega^*)$ on $]-\pi,\pi]$ by

(10.3) $$S^*_n(x;f;\omega^*) = \frac{1}{\pi}(pv)\int_{\omega^*} \frac{e^{-int} f^o(t)}{x-t} dt , \quad x \in]-\pi,\pi] ,$$

i.e. $S^*_n(x;f;\omega^*)$ is the Hilbert transform of $e^{-inx} f^o(x) \chi_{\omega^*}(x)$.

Since f is supposed to be a real valued function we immediately get

$$\overline{S^*_{-n}(x;f;\omega^*)} = S^*_n(x;f;\omega^*) , \quad n \in Z ,$$

where the bar denotes complex conjugation. Especially,

$$|S^*_{-n}(x;f;\omega^*)| = |S^*_n(x;f;\omega^*)| , \quad n \in Z .$$

The operator $M^* : L^p(]-\pi,\pi]) \to \mathfrak{M}$ is now defined by

(10.4) $$M^*f(x) = \sup_{n\in Z} |S^*_n(x;f;\omega^*_{-1})| = \sup_{n\in N_0} |S^*_n(x;f;\omega^*_{-1})| .$$

We note that M^* is only defined by means of $\omega^*_{-1} =]-4\pi, 4\pi]$, but in the estimation of M^* , however, we shall also need $S^*_n(x;f;\omega^*)$ for general ω^* .

Let $D_n(y)$, $n \in Z$, denote the *Dirichlet kernel* defined by

$$D_n(y) = \begin{cases} \dfrac{\sin(n+\frac{1}{2})y}{2\sin\frac{y}{2}} & \text{for } y \in [-\pi,\pi]\setminus\{0\} \\[2ex] n+\frac{1}{2} & \text{for } y = 0 \,. \end{cases}$$

Then it is a classical result (which is easy to verify) that for $n \in N_o$

$$S_n(x;f) = \sum_{k=-n}^{n} c_k e^{ikx} = \frac{1}{\pi} \int\limits_{|x-t|<\pi} f^0(t)\, D_n(x-t)dt \;.$$

In order to prove (10.2) we shall try to estimate $|S_n(x;f)|$ by means of $\|f\|_1$ $[\leq (2\pi)^{1-\frac{1}{p}} \|f\|_p]$ and $|S^*_n(x;f;\omega^*_{-1})|$. To do so we introduce another kernel $F_n(y)$, $n \in Z$, by

$$F_n(y) = \begin{cases} \dfrac{\sin ny}{y} & \text{for } y \in [-\pi,\pi]\setminus\{0\} \,, \\[2ex] n & \text{for } y = 0 \,. \end{cases}$$

Then for almost every $x \in]-\pi,\pi]$,

$$(10.5)\quad \frac{1}{\pi} \int\limits_{|x-t|<\pi} f^0(t)\, F_n(x-t)dt = \frac{1}{\pi}(pv) \int\limits_{|x-t|<\pi} \frac{e^{in(x-t)} - e^{-in(x-t)}}{2\,i\,(x-t)} f^0(t)dt$$

$$= \frac{e^{inx}}{2\pi\, i}(pv) \int\limits_{|x-t|<\pi} \frac{e^{-int} f^0(t)}{x-t}dt - \frac{e^{-inx}}{2\pi i}(pv) \int\limits_{|x-t|<\pi} \frac{e^{int} f^0(t)}{x-t}dt \;.$$

Now for all $m \in Z$,

$$\frac{1}{\pi}(pv) \int\limits_{|x-t|<\pi} \frac{e^{-imt} f^0(t)}{x-t}dt = S^*_m(x;f;\omega^*_{-1}) - \frac{1}{\pi} \int\limits_{\substack{|x-t|\geq\pi \\ |t|<4\pi}} \frac{e^{-imt} f^0(t)}{x-t}dt \;,$$

so from (10.5) we get the estimate

$$(10.6)\qquad \Big|\frac{1}{\pi}\int_{|x-t|<\pi} f^{o}(t)\,F_n(x-t)dt\Big| \leqq |S_n^*(x;f;\omega_{-1}^*)| + \frac{4}{\pi^2}\|f\|_1 .$$

Lemma 10.1. *For all* $y\in[-\pi,\pi]$ *and all* $n\in Z$ *we have*

$$|D_n(y)-F_n(y)| < 1 .$$

Proof. We introduce the function $g(y)$ by

$$g(y) = \begin{cases} \frac{1}{2}\cot\frac{y}{2}-\frac{1}{y} & \text{for } y\in[-\pi,\pi]\setminus\{0\}, \\ 0 & \text{for } y=0 . \end{cases}$$

It is easy to check that g is continuous and decreasing. Especially, $|g(y)| \leqq \frac{1}{\pi}$ for $y\in[-\pi,\pi]$. For $y=0$ the lemma is trivial. For $y\neq 0$ we get

$$D_n(y) = \frac{\sin(n+\frac{1}{2})y}{2\sin\frac{y}{2}} = \frac{1}{2}\sin ny\cdot\cot\frac{y}{2}+\frac{1}{2}\cos ny$$

$$= \sin ny\cdot\left\{\frac{1}{2}\cot\frac{y}{2}-\frac{1}{y}\right\}+\frac{\sin ny}{y}+\frac{1}{2}\cos ny$$

$$= g(y)\cdot\sin ny + F_n(y)+\frac{1}{2}\cos ny ,$$

so

$$|D_n(y)-F_n(y)| \leqq |g(y)|+\frac{1}{2} \leqq \frac{1}{\pi}+\frac{1}{2} < 1 . \qquad \square$$

Theorem 10.2. *If* $f\in L^p(]-\pi,\pi])$, $p\in\,]1,+\infty]$, *then*

$$\|Mf\|_p \leqq 5\|f\|_p + \|M^*f\|_p .$$

Proof. For every $n\in N_o$ and every $x\in\,]-\pi,\pi]$ we have

$$S_n(x;f) = \frac{1}{\pi}\int_{|x-t|<\pi} f^{o}(t)\,D_n(x-t)dt$$

$$= \frac{1}{\pi}\int_{|x-t|<\pi} f^{o}(t)\{D_n(x-t)-F_n(x-t)\}dt + \frac{1}{\pi}\int_{|x-t|<\pi} f^{o}(t)\,F_n(x-t)dt .$$

Using lemma 10.1 and (10.6) we thus get the following inequality for almost every $x \in]-\pi, \pi]$,

$$|S_n(x;f)| \leq \frac{1}{\pi} \cdot 1 \cdot \|f\|_1 + |S_n^*(x;f;\omega_{-1}^*)| + \frac{4}{\pi^2}\|f\|_1$$

$$\leq \frac{1}{\pi}(1+\frac{4}{\pi}) \cdot (2\pi)^{1-\frac{1}{p}} \|f\|_p + |S_n^*(x;f;\omega_{-1}^*)| ,$$

so

$$Mf(x) = \sup_{n \in N_o} |S_n(x;f)| \leq \frac{1}{\pi}(1+\frac{4}{\pi}) \cdot (2\pi)^{1-\frac{1}{p}} \|f\|_p + M^*f(x) , \quad \text{a.e.},$$

from which we get, using Minkowski's inequality,

$$\|Mf\|_p \leq \frac{1}{\pi}(1+\frac{4}{\pi}) \cdot (2\pi)^{1-\frac{1}{p}} \cdot (2\pi)^{\frac{1}{p}} \|f\|_p + \|M^*f\|_p$$

$$= 2(1+\frac{4}{\pi})\|f\|_p + \|M^*f\|_p \leq 5\|f\|_p + \|M^*f\|_p . \qquad \square$$

We shall now give some estimates for $S_n^*(x;f;\omega^*)$, where $\omega^* = \omega_{j\nu} \cup \omega_{j+1,\nu}$ is any smoothing interval composed of two neighbouring dyadic intervals, $\omega_{j\nu}$ and $\omega_{j+1,\nu}$, each of length $2\pi \cdot 2^{-\nu}$. Let $f \in L^1(]-\pi,\pi])$. We shall in this section consider

$$S_n^*(x;f;\omega^*) = \frac{1}{\pi}(pv)\int_{\omega^*} \frac{e^{-int} f^o(t)}{x-t} dt$$

for different values of n. Recall that $\overline{S_{-n}^*(x;f;\omega^*)} = S_n^*(x,f,\omega^*)$, so in the following lemma we shall only consider $n, n_o \in N_o$.

<u>Lemma 10.3.</u> *Let $n, n_o \in N$ and suppose that there exists an $m_o \in N_o$, such that $n_o = 2^{\nu+1} m_o$. If $|n-n_o| \leq 2^{\nu+1}$ then*

$$|e^{inx} S_n^*(x;f;\omega^*) - e^{in_o x} S_{n_o}^*(x;f;\omega^*)|$$

$$\leq c_4 \cdot \max\left\{C_{\psi(n_o;\omega_{j\nu})}(\omega_{j\nu};f) \; ; \; C_{\psi(n_o;\omega_{j+1,\nu})}(\omega_{j+1,\nu};f)\right\} ,$$

where $c_4 \in R_+$ *is a constant independent of* n, n_0, m_0 *and* ν *and* ω^* *and* f .

Proof. We note that [cf. (9.1)] we have $\psi(n_0;\omega_{j\nu}) = \psi(n_0;\omega_{j+1,\nu}) = 2m_0$. First we consider the function $\Phi(u)$ defined by

$$\Phi(u) = \begin{cases} \dfrac{e^{iu}-1}{u} & \text{for } u \in R\setminus\{0\} , \\ i & \text{for } u = 0 . \end{cases}$$

Then $\Phi \in C^\infty(R)$, and $\Phi(u) \to 0$ and $\Phi''(u) \to 0$ for $u \to +\infty$ or $u \to -\infty$, so $\|\Phi\|_\infty$, $\|\Phi''\|_\infty \le c_1'$ for some constant $c_1' \in R_+$.

Now, let $\varphi(t) = \frac{1}{t}\left\{\exp(i\{n-n_0\}t) - 1\right\} = (n-n_0)\,\Phi(\{n-n_0\}t)$ with trivial modifications for $t = 0$. From the above follows that

$$\|\varphi\|_\infty = |n-n_0| \cdot \|\Phi\|_\infty \le 2\cdot 2^\nu c_1' \ , \ \|\varphi''\|_\infty = |n-n_0|^3 \cdot \|\Phi''\|_\infty \le 8\cdot 2^{3\nu}\cdot c_1' \ .$$

Let us consider $\omega_{j\nu}$. Using lemma 9.4 we have

$$\varphi(t) = \sum_{\mu\in Z} \gamma_\mu \exp\left(i\,2^\nu\cdot\frac{\mu}{3}t\right) , \quad t \in \omega_{j\nu} + \{x\} ,$$

where

(10.7)
$$\begin{aligned}(1+\mu^2)|\gamma_\mu| &\le c_2\cdot\left\{\max_{\omega_{j\nu}}|\varphi| + 2^{-2\nu}\max_{\omega_{j\nu}}|\varphi''|\right\} \\ &\le c_2\cdot\left\{2\cdot 2^\nu c_1' + 8\cdot 2^\nu\cdot c_1'\right\} = c_2'\cdot 2^\nu .\end{aligned}$$

Then we have the following computations

$$\Delta_j = e^{inx}\int_{\omega_{j\nu}} \frac{e^{-int} f^o(t)}{x-t}\,dt - e^{in_0 x}\int_{\omega_{j\nu}} \frac{e^{-in_0 t} f^o(t)}{x-t}\,dt$$

$$= \int_{\omega_{j\nu}} \left\{e^{in(x-t)} - e^{in_0(x-t)}\right\} \frac{f^o(t)}{x-t}\,dt$$

$$= \int_{\omega_{j\nu}} e^{in_0(x-t)}\cdot\varphi(x-t)f^o(t)\,dt = \sum_{\mu\in Z}\gamma_\mu\int_{\omega_{j\nu}} \exp\left\{i(n_0 + 2^\nu\cdot\tfrac{\mu}{3})(x-t)\right\}f^o(t)\,dt$$

$$= \sum_{\mu\in Z} \gamma_\mu \exp\left\{i(n_0 + 2^\nu\cdot\tfrac{\mu}{3})x\right\} \int_{\omega_{j\nu}} \exp\left\{-i\, 2^\nu(2m_0+\tfrac{\mu}{3})t\right\} \cdot f^0(t)dt$$

$$= \sum_{\mu\in Z} \gamma_\mu \cdot \exp\left\{i(n_0 + 2^\nu\cdot\tfrac{\mu}{3})x\right\} \cdot c_{2m_0+\frac{\mu}{3}}(\omega_{j\nu};f) \cdot m(\omega_{j\nu}) ,$$

where we have used (9.3). Hence, according to (10.7),

$$(10.8) \qquad |\Delta_j| \le m(\omega_{j\nu}) \sum_{\mu\in Z} |\gamma_\mu| \cdot |c_{2m_0+\frac{\mu}{3}}(\omega_{j\nu};f)|$$

$$\le m(\omega_{j\nu}) \cdot \sum_{\mu\in Z} \frac{c_2'\cdot 2^\nu}{1+\mu^2} \cdot |c_{2m_0+\frac{\mu}{3}}(\omega_{j\nu};f)|$$

$$= 2\pi \cdot 10\, c_2' \cdot C_{2m_0}(\omega_{j\nu};f) = 20\pi \cdot c_2' \cdot C_{2m_0}(\omega_{j\nu};f) .$$

Similarly, if

$$\Delta_{j+1} = e^{inx} \int_{\omega_{j+1,\nu}} \frac{e^{-int} f^0(t)}{x-t} dt - e^{in_0 x} \int_{\omega_{j+1,\nu}} \frac{e^{-in_0 t} f^0(t)}{x-t} dt ,$$

then also

$$|\Delta_{j+1}| \le 20\pi \cdot c_2' \cdot C_{2m_0}(\omega_{j+1,\nu};f)$$

and therefore

$$|e^{inx} S_n^*(x;f;\omega^*) - e^{in_0 x} S_{n_0}^*(x;f;\omega^*)| = \frac{1}{\pi} |\Delta_j + \Delta_{j+1}|$$

$$\le 40\, c_2' \cdot \max\left\{C_{2m_0}(\omega_{j\nu};f) ;\; C_{2m_0}(\omega_{j+1,\nu};f)\right\} ,$$

and the lemma is proved with $c_4 = 40\, c_2'$. □

Remark 10.4. A careful examination of the function Φ above gives that we may choose $c_4 = 13{,}500$.

Remark 10.5. For later use we note that (10.8) is still valid, if $\omega_{j\nu}$ is replaced by one of its dyadic subintervals from level $\nu+1$. If $\omega_{k,\nu+1}$ is one of these subintervals, then

$$(10.9)\qquad |e^{inx}\int_{\omega_{k,\nu+1}} \frac{e^{-int} f^o(t)}{x-t}\,dt - e^{in_o x}\int_{\omega_{k,\nu+1}} \frac{e^{-in_o t} f^o(t)}{x-t}\,dt|$$

$$\le 20\pi \cdot c_2' \cdot C_{2m_o}(\omega_{k,\nu+1};f) .$$

Corollary 10.6. *Let* $\omega^* = \omega_{j\nu} \cup \omega_{j+1,\nu}$ *and* $n_o = m_o 2^{\nu+1}$ *for a suitable* $m_o \in \mathbb{N}_o$. *Let* $F \subseteq]-\pi,\pi]$ *be a measurable set. If* $|n-n_o| \leqq 2^{\nu+1}$, *then*

$$\Big||S_n^*(x;\chi_F;\omega^*)| - |S_{n_o}^*(x;\chi_F;\omega^*)|\Big| \leqq c_4 \cdot \max\Big\{C_{2m_o}(\omega_{j\nu};\chi_F) \; ; \; C_{2m_o}(\omega_{j+1,\nu};\chi_F)\Big\} ,$$

where $c_4 \in \mathbb{R}_+$ *is the constant introduced in lemma 10.3.*

Proof. The corollary follows immediately from lemma 10.3, as

$$\Big||S_n^*(x;\chi_F;\omega^*)| - |S_{n_o}^*(x;\chi_F;\omega^*)|\Big| = \Big||e^{inx} S_n^*(x;\chi_F;\omega^*)| - |e^{in_o x} S_{n_o}^*(x;\chi_F;\omega^*)|\Big|$$

$$\leqq |e^{inx} S_n^*(x;\chi_F;\omega^*) - e^{in_o x} S_{n_o}^*(x;\chi_F;\omega^*)| .$$

□

Lemma 10.7. *Suppose that* $n_o = 2^{\nu+1} m_o$ *for a suitable* $m_o \in \mathbb{N}$. *If* $|n-n_o| \le 2^{\nu+1}$, *then*

$$|e^{inx} S_n^*(x;f;\omega^*) - e^{in_o x} S_{n_o}^*(x;f;\omega^*)| \leqq c_5 \cdot C^*_{\psi^*(n_o;\omega^*)}(\omega^*;f) = c_5 \cdot C^*_{m_o}(\omega^*;f) .$$

Proof. Let $\omega^* = \omega_1 \cup \omega_2 \cup \omega_3 \cup \omega_4$, where $4m(\omega_j) = m(\omega^*) = 4 \cdot 2\pi \cdot 2^{-(\nu+1)}$, $j=1,\dots,4$, i.e. ω_j belongs to level $\nu+1$, $j=1,\dots,4$. Then by (9.10),

$$C^*_{m_o}(\omega^*;f) = \max\Big\{C_{2m_o}(\omega_j;f) \mid j=1,\dots,4\Big\} ,$$

so we get by using (10.9)

$$\left|e^{inx} S_n^*(x;f;\omega^*) - e^{in_o x} S_{n_o}^*(x;f;\omega^*)\right| \leqq 4 \cdot 20\, c_2' \cdot C_{m_o}^*(\omega^*;f) \ . \qquad \square$$

Remark 10.8. But using the exact value of c_2' we get $c_5 = 80\, c_2' \leqq 27{,}000$.

Lemma 10.9. *Let* ω^* *be a smoothing interval, and let* $m, n \in \mathbb{Z}$. *If* x *belongs to the middle half of* ω^* , *then*

$$|S_n^*(x;e^{imx};\omega^*| \leqq 5 \ .$$

Proof. Let $\overline{\omega^*}$ be the biggest subinterval of ω^* of the form $\overline{\omega^*} =]x-a\,,\,x+a]$. Then, (cf. the proof of lemma 8.2),

$$|S_n^*(x;e^{imx};\omega^*)| = \left|\frac{1}{\pi}(pv)\int_{\omega^*}\frac{e^{-int}e^{imt}}{x-t}dt\right|$$

$$\leqq \left|\frac{1}{\pi}(pv)\int_{\overline{\omega^*}}\frac{e^{i(m-n)t}}{x-t}dt\right| + 3\|\Theta(e^{i(m-n)x})\|_\infty$$

$$= \left|\frac{1}{\pi}(pv)\int_{\overline{\omega^*}}\frac{e^{i(m-n)t}}{x-t}dt\right| + 3$$

$$= \left|\frac{1}{\pi}e^{i(m-n)x}(pv)\int_{\overline{\omega^*}}\frac{e^{i(m-n)(t-x)}}{x-t}dt\right| + 3$$

$$= \left|\frac{1}{\pi}(pv)\int_{-a}^{a}\frac{e^{i(m-n)t}}{-t}dt\right| + 3 = \left|\frac{1}{\pi}(pv)\int_{-a}^{a}\frac{\sin(m-n)t}{t}\right| + 3$$

$$= \left|\frac{2}{\pi}\lim_{\varepsilon\to 0+}\int_{\varepsilon}^{a(m-n)}\frac{\sin u}{u}du\right| + 3 \leqq \frac{2}{\pi}\cdot\pi + 3 = 5 \ . \qquad \square$$

CHAPTER IV.

In this chapter we shall prove theorem 4.2, which remains of the proof of the Carleson-Hunt theorem. (This final proof is given in § 18.) We need, however, some technical results, before this can be done. As mentioned in the introduction to § 10 we shall prove that the operator M^* defined by (10.4) is of restricted type p for all $p \in]1,+\infty[$, since we then can use interpolation theorems from chapter I. Our goal is therefore to prove the existence of a constant $B_p \in R_+$ for $p \in]1,+\infty[$, such that

$$\text{(IV.1)} \qquad m(\{x \in]-\pi,\pi] \mid M^*\chi_F(x) > y\}) \leqq B_p^p\, y^{-p}\, m(F)$$

for every $y \in R_+$ and every measurable set $F \subseteqq]-\pi,\pi]$. This is done by constructing an exceptional set $E_N \subseteqq]-4\pi,4\pi]$ for each $N \in N$, such that

$$\text{(IV.2)} \qquad |S_n^*(x;\chi_F;\omega_{-1}^*)| \leqq y \quad \text{for} \quad x \in]-\pi,\pi] \setminus E_N \quad \text{and} \quad |n| \leqq N$$

and

$$\text{(IV.3)} \qquad m(E_N) \leqq B_p^p\, y^{-p}\, m(F) \; ,$$

since (IV.1) follows from (IV.2) and (IV.3). This exceptional set E_N is the union of some sets S^* and V_L^* (introduced in § 11), Y^* and X^* (introduced in § 12) and T^* and U^* (introduced in § 15). In each case we prove an estimate similar to (IV.3). In order to define X^* we have to introduce some auxiliary functions $P_k(x;\omega)$ in § 12 and prove some estimates for these in § 13. We shall use these functions to define some index sets G_k^* and $\tilde{G}$, which will be used for defining a splitting $\Omega(p^*;k)$ of the interval ω^* into disjoint subintervals, on which $C^*_{\psi^*(n;\omega^*)}(\omega^*;\chi_F^0)$ behaves fairly well. In § 17 we give the final estimate for $S_n^*(x;\chi_F;\omega_{-1}^*)$, which is needed in the proof of M^* being of restricted type p , but before that we have also given (in § 16) a fairly technical proof of some estimates for elements not contained in G_{kL}^* .

As $S_n^*(x;\chi_F;\omega_{-1}^*) = 0$ and thus $M^*\chi_F = 0$ if F is a null-set, we shall throughout §§ 11-17 assume that we are given $N \in N$, $p \in]1,+\infty[$, $y \in R_+$ and $F \subseteqq]-\pi,\pi]$ measurable, where $m(F) > 0$.

§ 11. Construction of the sets S^* and V_L^*.

Let ω denote any dyadic interval and ω^* any smoothing interval. For ω we denote by $\tilde{\omega}$ the set

(11.1) $$\tilde{\omega} = \{x \in]-2\pi, 2\pi] \mid \operatorname{dist}(x,\omega) \leqq 3 \cdot m(\omega)\} \setminus \{x_{\tilde{\omega}}\}$$

where $x_{\tilde{\omega}}$ is the left end point of $\tilde{\omega}$. Thus, normally $\tilde{\omega}$ is the union of seven neighbouring dyadic intervals from the same level with ω in the middle. If ω lies too close to one of the end points of the interval $]-2\pi, 2\pi]$, then $\tilde{\omega}$ is still the union of some neighbouring dyadic intervals, but the number of these may be less than seven.

We define

(11.2) $$S = \cup\left\{\omega \;\middle|\; y^{-p}\int_\omega \chi_F^o(x)\,dx \geqq m(\omega)\right\} \quad \text{and} \quad S^* = \bigcup_{\omega\subset S} \tilde{\omega}\ .$$

We note that $S=\emptyset$ and hence also $S^*=\emptyset$ if $y>1$. In fact, $y^{-p}\int_\omega \chi_F^o(x)\,dx < 1\cdot m(\omega)$ for $y>1$. In general we have the following estimate.

<u>Theorem 11.1.</u> *$m(S^*) \leqq 14 \cdot y^{-p} m(F)$.*

<u>Proof.</u> A direct computation gives

$$m(S^*) \leqq 7m(S) \leqq 7y^{-p}\int_{-2\pi}^{2\pi} \chi_F^o(x)\,dx = 14y^{-p}\int_{-\pi}^{\pi} \chi_F(x)\,dx = 14 \cdot y^{-p}\, m(F)\ . \qquad \square$$

In the sequel we shall only consider the dyadic intervals ω not contained in S. Let $\|\cdot\|_{p,\omega}$ denote the p-norm over ω.

Lemma 11.2. *If* $\omega \notin S$ *then*

$$\|\chi_F^o\|_{p,\omega} < \{m(\omega)\}^{1/p}\, y \tag{11.3}$$

and

$$C_n(\omega;\chi_F^o) < y \quad \textit{for all} \quad n \in \mathbb{Z}\ . \tag{11.4}$$

Proof. It follows from (11.2) that if $\omega \notin S$ then

$$y^{-p} \int_\omega \chi_F^o(x)\,dx < m(\omega) \ ,$$

and as $|\chi_F^o(x)|^p = \chi_F^o(x)$ we immediately get (11.3) .

Using (9.7), Hölder's inequality and (11.3) we finally get

$$0 \leqq C_n(\omega,\chi_F^o) \leqq \frac{1}{m(\omega)} \int_\omega |\chi_F^o(x)|\,dx \leqq \frac{1}{m(\omega)} \{m(\omega)\}^{1/q}\, \|\chi_F^o\|_{p,\omega} < y\ . \qquad \square$$

Lemma 11.3. *Suppose that* $\omega^* \notin S^*$ *and that* $\chi_F^o \cdot \chi_{\omega^*} \not\equiv 0$. *For every* $n \in \mathbb{Z}$ *one can find a constant* $k \in \mathbb{N}$ *such that*

$$2^{-k} \cdot y \leqq C_n^*(\omega^*;\chi_F^o) < 2 \cdot 2^{-k}\, y\ . \tag{11.5}$$

Proof. It follows from the equivalence of (9.11) and (9.12), and the definition (9.9) of $C_n^*(\omega^*;\chi_F)$ that $C_n^*(\omega^*;\chi_F) > 0$ for all $n \in \mathbb{Z}$. It is obvious that there exists an integer $k \in \mathbb{Z}$, such that (11.5) is satisfied. We have to prove that $k \in \mathbb{N}$.

From the condition $\omega^* \notin S^*$ follows that $\omega \notin S$ for all of the four subintervals $\omega \subset \omega^*$ for which $4m(\omega) = m(\omega^*)$. Hence $C_n(\omega;\chi_F^o) < y$ for each of these subintervals according to lemma 11.2 and then also

$$0 < C_n^*(\omega^*;\chi_F^o) = \max\left\{C_N(\omega;\chi_F^o) \mid \omega \subset \omega^*\ ,\ 4m(\omega) = m(\omega^*)\right\} < y\ ,$$

and (11.5) becomes trivial with $k \in \mathbb{N}$. $\square$

<u>Lemma 11.4.</u> *If* $\omega^* \notin S^*$ *then for every* $p \in]1,+\infty[$ *there exists a positive integer* $L = L(p) \in \mathbb{N}$ *, such that*

$$2^{-k} y \leqq C_n^*(\omega^*;\chi_F^o) \Rightarrow y^{p/2} \leqq 2^{Lk/4} y .$$

We may choose

$$(11.6) \qquad L = L(p) = \begin{cases} \left[\frac{2}{p-1}\right] & \text{for } 1 < p < \frac{51}{50} \\ 100 & \text{for } \frac{51}{50} \leq p \leqq 50 \\ [2p] & \text{for } 50 < p < +\infty \end{cases} .$$

<u>Proof.</u> If $\omega^* \notin S^*$ then $\omega \notin S$ for each of the four subintervals $\omega \subset \omega^*$ for which $4m(\omega) = m(\omega^*)$. Thus

$$C_n(\omega;\chi_F^o) \leqq \frac{1}{m(\omega)} \int_\omega \chi_F^o(x)dx < y^p ,$$

where we have used (9.7) and the definition (11.2) of S , so

$$(11.7) \quad 2^{-k} y \leqq C_n^*(\omega^*;\chi_F^o) = \max\left\{C_n(\omega;\chi_F^o) \mid \omega \subset \omega^* , 4m(\omega) = m(\omega^*)\right\} < y^p .$$

From (11.7) we conclude that $y > 2^{-k/(p-1)}$.

(a) If $1 < p \leqq 2$, then $\frac{p}{2} - 1 \leqq 0$ and so

$$y^{\frac{p}{2}-1} \leqq 2^{-\frac{k}{p-1}\left(\frac{p-2}{2}\right)} = 2^{\frac{2-p}{2(p-1)}k} ,$$

so the lemma is proved in this case, if $L(p) \geqq 2 \cdot \frac{2-p}{p-1}$, say

$$L(p) = \max\left\{\left[\frac{2}{p-1}\right] , 100\right\} , \qquad 1 < p \leqq 2 .$$

(b) If $2 < p < +\infty$ we use that

$$C_n(\omega;\chi_F^o) \leqq \frac{1}{m(\omega)} \int_\omega \chi_F^o(x)dx \leqq 1 ,$$

so $2^{-k} y \leqq C_n^*(\omega;\chi_F^o) \leqq 1$, from which follows that $y \leqq 2^k$. Since $\frac{p}{2} - 1 \geqq 0$ we get

$$y^{\frac{p}{2}-1} \leqq 2^{\frac{p-2}{2}k} ,$$

so in this case we may choose $L(p) \geqq 2(p-2)$, say $L(p) = \max\{[2p] , 100\}$.

□

It is for technical reasons in the sequel that $L(p)$ has been chosen as a positive integer $\geqq 100$. A closer examination of the places where $L(p)$ is applied would reveal that it suffices with $L(p) \geqq 36$.

We define

$$V_L^* = \{x \in \omega_{-1}^* \mid \hat{H}\, \chi_F^o(x) > L\, y\} \tag{11.8}$$

where $L = L(p)$ is the constant defined in (11.6), and where $\hat{H}\chi_F^o$ is the modified maximal Hilbert transform of χ_F^o with respect to ω_{-1}^* .

<u>Theorem 11.5.</u> $\quad m(V_L^*) \leqq 4\left(\frac{A_p^*}{L}\right)^p y^{-p} \cdot m(F)$.

<u>Proof.</u> From theorem 8.3 follows that $\hat{H}$ is of type p for all $p \in]1,+\infty[$, i.e.

$$\|\hat{H}f\|_p \leqq A_p^* \|f\|_p ,$$

and hence also of weak type p [cf. (1.5)] , and as [cf. (1.2)] $m(V_L^*) = \lambda_{\hat{H}\chi_F^o}(Ly)$ we get

$$m(V_L^*) \leqq (A_p^*)^p \cdot (Ly)^{-p} \cdot \|\chi_F^o\|_{p,\omega_{-1}^*}^p = 4\left(\frac{A_p^*}{L(p)}\right)^p \cdot y^{-p} \cdot m(F) . \qquad \square$$

§ 12. Construction of the $P_k(x;\omega)$-functions and the sets G_k and Y^* and X_k^* and X^*.

In this section we shall define the set where the Fourier coefficients of χ_F - roughly speaking - are fairly large and prove that this set satisfies an estimate of the same type as S^* in lemma 11.1.

First we consider ω_{r0} , $r = 1,2,$ where $\omega_{10} =]-2\pi, 0]$ and $\omega_{20} =]0,2\pi]$, cf. (8.1). If $f \in L^2(]-\pi,\pi]$ we have a Fourier expansion for f , which is convergent in $L^2(]-\pi,\pi])$, hence

$$\sum_{n\in Z} |c_n(\omega_{r0};f)|^2 < +\infty \ .$$

Thus, given any positive constant $\alpha \in R_+$, then $|c_n(\omega_{r0};f)| \geqq \alpha$ for only a finite number of the Fourier coefficients, if any.

Let $k \in N$ be given. We define

(12.1) $$G_k(\omega_{r0}) = \{(n;\omega_{r0}) \ \big| \ n \in Z \ , \ |c_n(\omega_{r0};\chi_F^0)| \geqq 2^{-k} y^{p/2}\} \ .$$

Then $G_k(\omega_{r0})$ has a finite number of elements, if any. Furthermore, as $|c_n(\omega_{r0};\chi_F^0)| = |c_{-n}(\omega_{r0};\chi_F^0)|$, it follows that $(n,\omega_{r0}) \in G_k(\omega_{r0})$ if and only if $(-n,\omega_{r0}) \in G_k(\omega_{r0})$.

Having thus defined $G_k(\omega_{r0})$ we introduce the functions $R_k(x;\omega_{r0})$ and $P_k(x;\omega_{r0})$ [polynomials in e^{ix} and e^{-ix}] by

$$R_k(x;\omega_{r0}) = \sum_{(n,\omega_{r0}) \in G_k(\omega_{r0})} c_n(\omega_{r0};\chi_F^0)\, e^{inx} \ , \quad x \in]-2\pi, 2\pi] \ ,$$

and

$$P_k(x;\omega_{r0}) = R_k(x;\omega_{r0}) \ , \quad x \in]-2\pi, 2\pi] \ .$$

If $G_k(\omega_{r0}) = \emptyset$ then of course $R_k(x;\omega_{r0}) \equiv 0$.

We see that we in the definition of $R_k(x;\omega_{r0})$ just have picked up the terms in the Fourier expansion of χ_F^o which are large in some sense.

Next we consider the intervals ω_{s1} from level $\nu = 1$. Let $\omega_{s1} \subset \omega_{r0}$. Then we expand the function $\chi_F^o - P_k(\cdot;\omega_{r0})$ as a Fourier series over ω_{s1} and define

$$G_k(\omega_{s1}) = \left\{(n,\omega_{s1}) \;\middle|\; n \in Z \,,\ |c_n(\omega_{s1};\, \chi_F^o - P_k(\cdot;\omega_{r0}))| \geqq 2^{-k} y^{p/2}\right\} .$$

It is obvious that $G_k(\omega_{s1})$ only contains a finite number of elements, if any, so we may define

$$R_k(x;\omega_{s1}) = \sum_{(n,\omega_{s1}) \in G_k(\omega_{s1})} c_n(\omega_{s1};\, \chi_F^o - P_k(\cdot;\omega_{r0}))e^{i2nx} \,, \quad x \in \,]-2\pi,2\pi]$$

and

$$P_k(x;\omega_{s1}) = P_k(x;\omega_{r0}) + R_k(x;\omega_{s1}) = R_k(x;\omega_{r0}) + R_k(x;\omega_{s1}) \,, \quad x \in \,]-2\pi,2\pi] .$$

In general, suppose that $\omega_{j\nu} \subset \omega_{\ell,\nu-1}$, and suppose that the function $P_k(x;\omega_{\ell,\nu-1})$ has been constructed. We define

$$(12.2) \quad G_k(\omega_{j\nu}) = \left\{(n,\omega_{j\nu}) \;\middle|\; n \in Z \,, |c_n(\omega_{j\nu};\, \chi_F^o - P_k(\cdot;\omega_{\ell,\nu-1}))| \geqq 2^{-k} y^{p/2}\right\} ,$$

which only contains a finite number of elements, if any, and

$$(12.3) \quad R_k(x;\omega_{j\nu}) = \sum_{(n,\omega_{j\nu}) \in G_k(\omega_{j\nu})} c_n(\omega_{j\nu};\, \chi_F^o - P_k(\cdot;\omega_{\ell,\nu-1}))e^{i2^\nu nx} \,, \quad x \in \,]-2\pi,2\pi] ,$$

and

$$(12.4) \qquad P_k(x;\omega_{j\nu}) = P_k(x;\omega_{\ell,\nu-1}) + R_k(x;\omega_{j\nu}) \,, \quad x \in \,]-2\pi,2\pi] .$$

If $\omega_{j\nu} \subset \omega_{\ell,\nu-1} \subset \ldots \subset \omega_{s1} \subset \omega_{r0}$, it is easy to see that

$$(12.5) \quad P_k(x;\omega_{j\nu}) = R_k(x;\omega_{r0}) + R_k(x;\omega_{s1}) + \ldots + R_k(x;\omega_{\ell,\nu-1}) + R_k(x;\omega_{j\nu}) .$$

In this way we continue, and we see that we for any $k \in N$ and any dyadic interval ω have a set $G_k(\omega)$ with a finite number of elements and two functions $P_k(x;\omega)$ and $R_k(x;\omega)$, which are polynomials in e^{ix} and e^{-ix}.

Remark 12.1. Note that by the construction of $R_k(x;\omega_{j\nu})$ in (12.3) we have removed all the large terms of the Fourier expansion of $\chi_F^o - P_k(\cdot;\omega_{\ell,\nu-1})$ over $\omega_{j\nu}$, so using (12.4) and (12.2) it follows that

$$|c_n(\omega_{j\nu};\chi_F^o - P_k(\cdot;\omega_{j\nu}))| < 2^{-k}y^{p/2} \quad \text{for all } n \in Z .$$

Remark 12.2. In general, the notation above is a little awkward, so we shall often write $a_n(\omega)$ for short instead of $c_n(\omega_{j\nu};\chi_F^o - P_k(\cdot;\omega_{\ell,\nu-1}))$, thus

$$R_k(x;\omega) = \sum_{(n,\omega)\in G_k(\omega)} a_n(\omega)e^{i2\pi\, m(\omega)^{-1}nx}, \quad x \in]-2\pi,2\pi] ,$$

where of course $a_n(\omega)$ also depends upon k and F. Sometimes we shall even replace $a_n(\omega)e^{i2^k nx}$ by $a\,e^{i\lambda x}$. This simplification is of great help in the following. It should always be remembered, however, that if we consider a particular term $a\,e^{i\lambda x}$ from $P_k(x;\omega)$ we shall always assume that it originates from one and only one $R_k(x;\omega')$, where $\omega' \supseteq \omega$, so we never collect terms from different R_k-functions, although terms with the same λ may occur in more than one of the R_k-functions.

We now collect all the sets defined in (12.2) in the definition

$$(12.6) \qquad G_k = \bigcup_{\omega\subset]-2\pi,2\pi]} G_k(\omega) , \quad k \in N ,$$

which of course is a disjoint union. In particular, $(n,\omega) \in G_k$ defines a uniquely determined term $a_n(\omega)$, namely the coefficient of $e^{i2\pi\, m(\omega)^{-1}nx}$ in $R_k(x;\omega)$.

<u>Lemma 12.3.</u> $\sum_{(n,\omega)\in G_k} |a_n(\omega)|^2 \cdot m(\omega) \leqq 2m(F)$.

<u>Proof.</u> Consider any $\nu \in \mathbb{N}_0$ and assume that $\omega_{j\nu} \subset \omega_{\ell,\nu-1}$, where we for $\nu = 0$ put $\omega_{\ell,-1} = \omega_{-1} =]-2\pi, 2\pi]$ and also $P_k(x;\omega_{\ell-1}) = 0$. Using (12.4) we get

$$(12.7)\quad \int_{\omega_{j\nu}} |\chi_F^o(x) - P_k(x;\omega_{j\nu})|^2 dx = \int_{\omega_{j\nu}} |\chi_F^o(x) - P_k(x;\omega_{\ell,\nu-1}) - R_k(x;\omega_{j\nu})|^2 dx .$$

Using the Fourier expansion of $\chi_F^o(x) - P_k(x;\omega_{\ell,\nu-1})$ and $R_k(x;\omega_{j\nu})$ over $\omega_{j\nu}$, where the latter expansion is given by (12.3), we conclude that $R_k(x;\omega_{j\nu})$ and $\chi_F^o(x) - P_k(x;\omega_{\ell,\nu-1}) - R_k(x;\omega_{j\nu})$ are orthogonal to each other in $L^2(\omega_{j\nu})$, so

$$\int_{\omega_{j\nu}} |\chi_F^o(x) - P_k(x;\omega_{\ell,\nu-1}) - R_k(x;\omega_{j\nu})|^2 dx + \int_{\omega_{j\nu}} |R_k(x;\omega_{j\nu})|^2 dx$$

$$= \int_{\omega_{j\nu}} |\chi_F^o(x) - P_k(x;\omega_{\ell,\nu-1})|^2 dx \quad .$$

After a rearrangement of this equation we get by using (12.7) that

$$(12.8)\quad \sum_{(n,\omega_{j\nu})\in G_k(\omega_{j\nu})} |a_n(\omega_{j\nu})|^2 m(\omega_{j\nu}) = \int_{\omega_{j\nu}} |R_k(x;\omega_{j\nu})|^2 dx$$

$$= \int_{\omega_{j\nu}} |\chi_F^o(x) - P_k(x;\omega_{\ell,\nu-1})|^2 dx - \int_{\omega_{j\nu}} |\chi_F^o(x) - P_k(x;\omega_{j\nu})|^2 dx .$$

For $\nu = 0$ the first term on the right hand side of (12.8) is equal to $\int_{\omega_{r0}} |\chi_F^o(x) - 0|^2 dx = m(F)$, so if we use that each $\omega_{\ell,\nu-1}$ contains two intervals from level ν , we get by a summation of (12.8) over all ω from the same level ν that

$$\sum_{\substack{(n,\omega)\in G_k(\omega)\\ m(\omega)=2\pi\cdot 2^{-\nu}}} |a_n(\omega)|^2 m(\omega) = \sum_{m(\omega)=2\pi\cdot 2^{-(\nu-1)}} \int_\omega |\chi_F^o(x) - P_k(x;\omega)|^2 dx$$

$$- \sum_{m(\omega)=2\pi\cdot 2^{-\nu}} \int_\omega |\chi_F^o(x) - P_k(x;\omega)|^2 dx ,$$

or by iteration

$$\sum_{\substack{(n,\omega)\in G_k(\omega)\\ m(\omega)\geq 2\pi\cdot 2^{-\nu}}} |a_n(\omega)|^2 m(\omega) = \int_{-2\pi}^{2\pi} |\chi_F^o(x)|^2 dx$$

$$- \sum_{m(\omega)=2\pi\cdot 2^{-\nu}} \int_\omega |\chi_F^o(x) - P_k(x;\omega)|^2 dx \leq 2m(F) .$$

As this is true for all $\nu \in N_o$ we finally get for $\nu \to +\infty$ that

$$\sum_{(n,\omega)\in G_k} |a_n(\omega)|^2 m(\omega) \leq 2m(F) . \qquad \square$$

<u>Corollary 12.4.</u> $\sum_{(n,\omega)\in G_k} m(\omega) \leq 2^{2k+1} y^{-p} m(F)$.

<u>Proof.</u> If $(n,\omega)\in G_k$, then $|a_n(\omega)| \geq 2^{-k} y^{p/2}$ and so $1 \leq 2^{2k} y^{-p} |a_n(\omega)|^2$. From lemma 12.3 follows that

$$\sum_{(n,\omega)\in G_k} m(\omega) \leq 2^{2k}\cdot y^{-p} \sum_{(n,\omega)\in G_k} |a_n(\omega)|^2 m(\omega) \leq 2^{2k+1} y^{-p} m(F) . \qquad \square$$

We shall now introduce the exceptional sets X^* and Y^* . Using the sets G_k we define for each $k\in N$ a function $A_k(x)$ on $]-2\pi, 2\pi]$ by

(12.9) $$A_k(x) = \sum_{(n,\omega)\in G_k} |a_n(\omega)|^2 \chi_\omega(x) ,$$

and we define the set X_k by

(12.10) $$X_k = \{x \in\,]-2\pi, 2\pi] \mid A_k(x) > 2^k y^p\} .$$

If $x\in X_k$, one can find a *finite* subset I of the index set G_k, such that $\sum_{(n,\omega)\in I} |a_n(\omega)|^2 \chi_\omega(x) > 2^k y^p$, and as the left hand side is a step function, there exists a dyadic interval ω' such that $x\in\omega'$ and $A_k(z) > 2^k y^p$ for all $z\in\omega'$. Hence we have proved the existence of a dyadic interval

ω' , such that $x \in \omega' \subseteq X_k$, so X_k is a union of dyadic intervals. Using the notation from (11.1) we define

(12.11) $$X_k^* = \bigcup_{\omega \subset X_k} \tilde{\omega} \quad \text{and} \quad X^* = \bigcup_{k=1}^{+\infty} X_k^* .$$

Theorem 12.5. $m(X^*) \leq 14 \cdot y^{-p} m(F)$.

Proof. The theorem follows from a small computation:

$$m(X^*) \leq \sum_{k=1}^{+\infty} m(X_k^*) \leq 7 \sum_{k=1}^{+\infty} m(X_k)$$

$$\leq 7 \sum_{k=1}^{+\infty} \int_{-2\pi}^{2\pi} 2^{-k} y^{-p} A_k(x)\,dx = 7 \cdot y^{-p} \sum_{k=1}^{+\infty} 2^{-k} \sum_{(n,\omega) \in G_k} |a_n(\omega)|^2 m(\omega)$$

$$\leq 7 y^{-p} \sum_{k=1}^{+\infty} 2^{-k} \cdot 2m(F) = 14 y^{-p} m(F) ,$$

where we have used lemma 12.3. □

Let ω be any dyadic interval, say $\omega =]2\pi \cdot j \cdot 2^{-\nu} , 2\pi(j+1) \cdot 2^{-\nu}]$. By F_ω^k , $k \in \mathbb{N}$, we shall understand the set

$$F_\omega^k = \left(\{2\pi \cdot j \cdot 2^{-\nu}\} +]-2\pi \cdot 2^{-\nu-3k} , 2\pi \cdot 2^{-\nu-3k}]\right)$$

$$\cup \left(\{2\pi(j+1) \cdot 2^{-\nu}\} +]-2\pi \cdot 2^{-\nu-3k} , 2\pi \cdot 2^{-\nu-3k}]\right) ,$$

i.e. F_ω^k is the union of the two dyadic intervals from level $\nu + 3k$ having the left end point of ω as common end point, and the two dyadic intervals from level $\nu + 3k$ having the right end point of ω as common end point. Then especially $m(F_\omega^k) = 4 \cdot 2^{-3k} m(\omega)$. We define

(12.12) $$Y^* = \bigcup_{k=1}^{+\infty} \bigcup_{(n,\omega) \in G_k} F_\omega^k .$$

Theorem 12.6. $m(Y^*) \leq 8 \cdot y^{-p} m(F)$.

Proof. Using corollary 12.4 we get

$$m(Y^*) \leq \sum_{k=1}^{+\infty} \sum_{(n,\omega)\in G_k} m(F_\omega^k) = 4 \sum_{k=1}^{+\infty} 2^{-3k} \sum_{(n,\omega)\in G_k} m(\omega) \leq 4 \sum_{k=1}^{+\infty} 2^{-3k} \cdot 2^{2k+1} y^{-p} m(F)$$

$$= 8 \cdot y^{-p} m(F) . \quad \square$$

§ 13. Estimates of $P_k(x;\omega)$ and introduction of the index set G_k^* .

From the given measurable set F we have constructed the functions $P_k(x;\omega)$ and the sets X_k , where $P_k(x;\omega)$ is too large. These auxiliary functions $P_k(x;\omega)$ do not, however, define the terms $S_n^*(x;\chi_F^o;\omega_{-1}^*)$ associated with χ_F^o [cf. the definition (10.4) of M^*] , so we shall need some method to estimate those terms, which in some sense are not too far away from the terms of $P_k(x;\omega)$. This sense is defined by means of the size of the set ω and the corresponding factor λ in the exponent of each term $a\,e^{i\lambda x}$ from $P_k(x;\omega)$, so in general we shall consider pairs $(n,\omega) \in \mathcal{P}$ where $n \in Z$ and ω is a dyadic interval, and we shall try to define what is meant by $(n,\omega) \in \mathcal{P}$ being close to some element from G_k , i.e. the index set for the functions P_k .

As X^* and X_k already have been defined as exceptional sets we shall only be concerned with dyadic intervals ω not included in the set X_k defined by (12.10). First we need an estimate of $P_k(x;\omega)$ itself and also of the number of terms in $P_k(x;\omega)$.

Lemma 13.1. *If* $\omega \notin X_k$ *then*

(13.1) $\qquad P_k(x;\omega)$ *contains at most* 2^{3k} *terms;*

and

(13.2) $$|P_k(x;\omega)| \leq \sum_{j=1}^{J} |a_j| \leq 2^{2k} y^{p/2} \quad \textit{for } x \in]-2\pi, 2\pi] ,$$

where

(13.3) $$P_k(x;\omega) = \sum_{j=1}^{J} a_j \exp(i \lambda_j x) .$$

<u>Proof.</u> From (12.2) – (12.4) follows that $|a_j| \geqq 2^{-k} y^{p/2}$ for $1 \leqq j \leqq J$, so $1 \leqq |a_j|^2 \, 2^{2k} y^{-p}$. Hence,

$$J \leqq \sum_{j=1}^{J} |a_j|^2 \cdot 2^{2k} y^{-p} = 2^{2k} \cdot y^{-p} \sum_{j=1}^{J} |a_j|^2 .$$

Choose any $x_o \in \omega \backslash X_k$. By (12.9) we have

$$\sum_{j=1}^{J} |a_j|^2 \leqq A_k(x_o) , \tag{13.4}$$

and by (12.10) we get

$$A_k(x_o) \leqq 2^k y^p , \tag{13.5}$$

so

$$J \leqq 2^{2k} y^{-p} A_k(x_o) \leqq 2^{3k} ,$$

proving (13.1).

Now, $1 \leqq |a_j| \cdot 2^k y^{-p/2}$ for $1 \leqq j \leqq J$ by construction, so from (13.3) – (13.5) follows, using any $x_o \in \omega \backslash X_k$ that

$$|P_k(x;\omega)| \leqq \sum_{j=1}^{J} 1 \cdot |a_j| \leqq 2^k \cdot y^{-p/2} \sum_{j=1}^{J} |a_j|^2 \leqq 2^k y^{-p/2} A_k(x_o) \leqq 2^{2k} y^{p/2} . \qquad \square$$

Let $\omega = \omega_{j\nu}$ be any dyadic interval from level ν not contained in X_k. If $a\, e^{i\lambda x}$ is any term from $P_k(x;\omega_{j\nu})$, then [cf. § 12] by construction $\bar{a}\, e^{-i\lambda x}$ is also a term from $P_k(x;\omega_{j\nu})$. This means that in the sequel we need only consider exponents $\lambda \in N_o$. Also by construction, there exists another dyadic interval $\omega_{\ell\mu} \supseteqq \omega_{j\nu}$, $\mu \leqq \nu$, such that $a\, e^{i\lambda x}$ is a term from $R_k(x;\omega_{\ell\mu})$. Hence there exists an integer $n \in N_o$, such that $\lambda = 2^{\mu} n$, i.e. [cf. (9.1)]

$$\psi(\lambda;\omega_{\ell\mu}) = n = \frac{\lambda}{2\pi} m(\omega_{\ell\mu})$$

and

$$(\psi(\lambda;\omega_{\ell\mu}), \omega_{\ell\mu}) = (n,\omega_{\ell\mu}) \in G_k .$$

If on the other hand $(n;\omega_{\ell\mu}) \in G_k$ then $P_k(x;\omega_{\ell\mu})$ contains a term $a\, e^{i\lambda x}$ where $\lambda = n \cdot 2^{\mu}$.

We shall now define the set $\tilde{G}_k$ of indices from $\mathcal{P}$ which in some sense are close to the set G_k associated with the $P_k(x;\omega)$-functions.

Definition 13.2. *By* $\tilde{G}_k^1$ *we shall understand the set of* $(n,\omega) \in \mathcal{P}$, $n \in N_o$ *and* $\omega \notin X_k$, *for which there exist a dyadic interval* $\omega' \supseteq \omega$ *and a term* $a'e^{i\lambda' x}$ *from* $R_k(x;\omega')$, *i.e.* $(\psi(\lambda';\omega'),\omega') \in G_k$, *such that*

$$|n - \psi(\lambda';\omega)| < 2^{10k} \tag{13.6}$$

and

$$m(\omega) > 2^{-10k}\, m(\omega') . \tag{13.7}$$

It should be noted that we write $\psi(\lambda';\omega)$ $(= [\psi(\lambda';\omega') \cdot \frac{m(\omega)}{m(\omega')}])$ in (13.6) Furthermore, if $\omega = \omega_{j\nu}$ and $\omega' = \omega_{\ell\mu}$ it follows from (13.7) that $\mu \leq \nu < \mu + 10k$, so we have in this case only $10k$ possible choices of the dyadic interval ω'. On the other hand, for given $(\psi(\lambda';\omega'),\omega') \in G_k$, i.e. for given λ' and ω' we can for each $\omega \subseteq \omega'$ satisfying (13.7) find at most $2 \cdot 2^{10k} - 1$ non-negative integers $n \in N_o$ fulfilling (13.6), so if $\tilde{G}_k(\lambda',\omega')$ denotes the set of all elements $(n,\omega) \in \tilde{G}_k^1$ associated with one particular pair (λ',ω'), then

$$\sum_{(n,\omega)\in\tilde{G}_k^1(\lambda',\omega')} m(\omega) = \sum_{\nu=\mu}^{\mu+10k-1} \sum_{\substack{(n,\omega)\in\tilde{G}_k^1(\lambda',\omega') \\ m(\omega)=2\pi\cdot 2^{-\nu}}} m(\omega) \leq \sum_{\nu=\mu}^{\mu+10k-1} (2\cdot 2^{10k}-1)\, m(\omega')$$

$$< 10 \cdot k \cdot 2 \cdot 2^{10k}\, m(\omega') < 20 \cdot 2^{11k}\, m(\omega') ,$$

where we have used that all different ω from the same level ν are disjoint and included in ω'. Hence,

$$\sum_{(n,\omega)\in\tilde{G}_k^1} m(\omega) = \sum_{(\psi(\lambda';\omega'),\omega')\in G_k} \; \sum_{(n,\omega)\in\tilde{G}_k^1(\lambda',\omega')} m(\omega) \tag{13.8}$$

$$< 20 \cdot 2^{11k} \sum_{(r,\omega')\in G_k} m(\omega') \leq 40 \cdot 2^{13k} \cdot y^{-p}\, m(F) ,$$

where we have used corollary 12.4.

When comparing elements from $\mathcal{P}$ with elements from G_k the condition $|n-\psi(\lambda';\omega)| < 2^{10k}$ could be satisfied for some $(\psi(\lambda';\omega'),\omega') \in G_k$, while ω' is too big compared with ω . In this case we shall search for another term $(\psi(\lambda'';\omega''),\omega'') \in G_k$, such that - roughly speaking - we have

$$2^{10k} \leqq |n-\psi(\lambda'';\omega)| < 2^{20k} , \tag{13.9}$$

i.e. we extend the range of the exponent, when the set ω is too small. This is the essential fact in the following definition, where we so to speak "approximate" with an exponent in the range $[0,2^{10k}[$ and with another one in the range $[2^{10k}, 2^{20k}[$. For technical reasons, however, we shall use another description.

Definition 13.3. *By $\tilde{G}_k^2$ we shall understand the set of $(n,\omega) \in \mathcal{P}$, $n \in N_0$ and $\omega \notin X_k$, for which there exist a dyadic interval ω' , such that $\omega \subseteqq \omega'$, and two terms $a' e^{i\lambda' x}$ and $a'' e^{i\lambda'' x}$ from $P_k(x;\omega')$, such that*

$$|n-\psi(\lambda';\omega)| < 2^{10k}$$

and

$$2^{-10k} \leqq |\lambda' - \lambda''| \cdot \frac{m(\omega)}{2\pi} < 2^{20k} . \tag{13.10}$$

Using the triangle inequality and that $\psi(\lambda'';\omega)$ is an integer it is easy to derive (13.9) from (13.10). We shall, however, not use this fact.

If $\omega \notin X_k$ it follows from lemma 13.1 that $P_k(x;\omega')$ contains at most 2^{3k} terms and hence at most 2^{6k} pairs of exponents (λ',λ'') as defined in definition 13.3. From (13.10) we get a substitute for condition (13.7) in definition 13.2, because it follows from (13.10) that given any pair (λ',λ'') of exponents the dyadic interval ω can only belong to at most 30k levels, say $\mu+1$, ... , $\mu+30k$, where μ also depends on (λ',λ'') . Using the same technique as in the derivation of (13.8), i.e. first considering all $(n,\omega) \in \tilde{G}_k^2$ originating from the same pair (λ',λ'') , where furthermore all ω belong to the same level, then collecting all the 30k levels and finally using corollary 12.4 once more, we get

$$(13.11) \qquad \sum_{(n,\omega)\in\tilde{G}_k^2} m(\omega) \leqq 120 \cdot 2^{19k} \cdot y^{-p}\, m(F) \ .$$

The extra factor 2^{6k} comes of course from the number of pairs (λ',λ'') .

We write for short

$$(13.12) \qquad \tilde{G}_k = \tilde{G}_k^1 \cup \tilde{G}_k^2 \ .$$

<u>Lemma 13.4.</u> $\displaystyle\sum_{(n,\omega)\in\tilde{G}_k} m(\omega) \leqq 125 \cdot 2^{19k} \cdot y^{-p}\, m(F)$.

<u>Proof.</u> As $k \in \mathbb{N}$, the lemma follows immediately from (13.8) and (13.11).

□

For each $k \in \mathbb{N}$ we define

$$(13.13) \qquad G_k^* = \{(n,\omega^*) \mid \exists\omega : (n,\omega) \in \tilde{G}_k \wedge \omega \subset \omega^* \wedge 4m(\omega) = m(\omega^*)\} \ ,$$

where ω^* as usual denotes a smoothing interval.

<u>Lemma 13.5.</u> $\displaystyle\sum_{(n,\omega^*)\in G_k^*} m(\omega^*) \leqq 500 \cdot 2^{19k}\, y^{-p}\, m(F)$.

<u>Proof.</u> This is a trivial consequence of lemma 13.4 and (13.13). □

<u>Remark 13.6.</u> If ω does not lie too close to either -2π or 2π there are exactly two smoothing intervals with ω in the middle half, such that $m(\omega^*) = 4m(\omega)$.

Furthermore, if $\omega^* \notin X_k^*$ [cf. § 12] and $(n,\omega^*) \notin G_k^*$, then $\omega \notin X_k$ and $(n,\omega) \notin \tilde{G}_k$ for each of the four intervals $\omega \subset \omega^*$, for which $4m(\omega) = m(\omega^*)$.

We note that $(n,\omega^*_{-1}) \in G^*_k$ if and only if $(n,\omega_{1,0}) \in \tilde{G}_k$ and $(n,\omega_{2,0}) \in \tilde{G}_k$, where we have used the periodicity of χ^0_F.

Suppose that $(n,\omega) \notin \tilde{G}_k$. Then

$$P_k(x;\omega) = Q^k_0(x;\omega) + Q^k_1(x;\omega) ,$$

where $Q^k_0(x;\omega)$ contains those terms $a'e^{i\lambda' x}$ from $P_k(x;\omega)$ for which

$$|n - \psi(\lambda';\omega)| < 2^{10k} ,$$

i.e. for which (13.6) is satisfied. As $(n,\omega) \notin \tilde{G}^2_k$, condition (13.10) cannot be fulfilled, so either $|\lambda' - \lambda''| \cdot \frac{m(\omega)}{2\pi} < 2^{-10k}$ or $|\lambda' - \lambda''| \cdot \frac{m(\omega)}{2\pi} > 2^{20k}$ for any two exponents λ' and λ'' occurring in $Q^k_0(x;\omega)$. But

$$|\lambda' - \lambda''| \cdot \frac{m(\omega)}{2\pi} = \left|\lambda' \cdot \frac{m(\omega)}{2\pi} - \lambda'' \cdot \frac{m(\omega)}{2\pi}\right| \leqq |\psi(\lambda';\omega) - \psi(\lambda'';\omega)| + 1$$

$$\leqq |n - \psi(\lambda';\omega)| + |n - \psi(\lambda'';\omega)| + 1 < 2 \cdot 2^{10k} + 1 < 2^{20k} ,$$

so the latter possibility is ruled out, and we get

$$|\lambda' - \lambda''| \cdot \frac{m(\omega)}{2\pi} < 2^{-10k} . \tag{13.14}$$

We may of course write

$$Q^k_0(x;\omega) = \sum_{j=1}^{I} a_j e^{i\lambda_j x} , \qquad P_k(x;\omega) = \sum_{j=1}^{J} a_j e^{i\lambda_j x} ,$$

where $I \leqq J$, so the terms with index $j = I+1, \ldots, J$ belong to $Q^k_1(x;\omega)$.

<u>Lemma 13.7.</u> *Let* λ *be any exponent occurring in* $Q^k_0(x;\omega)$. *Then there exists a constant* ρ, *such that*

$$|Q^k_0(x;\omega) - \rho e^{i\lambda x}| \leqq 23 \cdot 2^{-8k} \cdot y^{p/2} \qquad \text{for } x \in \tilde{\omega} .$$

Here $\tilde{\omega}$ *is defined by (11.1).*

Proof. Let x_o be the midpoint of ω and suppose that $\lambda = \lambda_1$. For $x \in \tilde{\omega}$ and $j = 1, \ldots, I$ we have the series expansion

$$\exp[i(\lambda_j - \lambda_1)(x - x_o)] = 1 + \sum_{n=1}^{+\infty} \frac{1}{n!} i^n[(\lambda_j - \lambda_1)(x - x_o)]^n .$$

Using (13.14) we get for $x \in \tilde{\omega}$

$$|\lambda_j - \lambda_1| \cdot |x - x_o| \leqq \frac{7}{2} \cdot |\lambda_j - \lambda_1| \cdot m(\omega) < \frac{7}{2} \cdot 2^{-10k} \cdot 2\pi < 22 \cdot 2^{-10k} ,$$

so

$$\text{(13.15)} \qquad |e^{i(\lambda_j - \lambda_1)(x - x_o)} - 1| < 22 \cdot 2^{-10k} \sum_{n=0}^{+\infty} \frac{1}{(n+1)!} \left[\frac{7}{2} \cdot 2^{-10k}\right]^n$$

$$< 23 \cdot 2^{-10k} , \quad k \in N .$$

Now,

$$Q_0^k(x;\omega) = \left\{\sum_{j=1}^{I} a_j e^{i(\lambda_j - \lambda_1)x_o}\right\} e^{i\lambda_1 x}$$

$$+ e^{i\lambda_1 x} \sum_{j=1}^{I} a_j e^{i(\lambda_j - \lambda_1)x_o} \left\{e^{i(\lambda_j - \lambda_1)(x - x_o)} - 1\right\} ,$$

so

$$\left|Q_0^k(x;\omega) - \left\{\sum_{j=1}^{I} a_j e^{i(\lambda_j - \lambda_1)x_o}\right\} e^{i\lambda_1 x}\right| \leqq \sum_{j=1}^{I} |a_j| \cdot \left|e^{i(\lambda_j - \lambda_1)(x - x_o)} - 1\right|$$

$$\leqq \sum_{j=1}^{I} |a_j| \cdot 23 \cdot 2^{-10k}$$

$$\leqq 23 \cdot 2^{-10k} \sum_{j=1}^{J} |a_j| \leqq 23 \cdot 2^{-10k} \cdot 2^{2k} \cdot y^{p/2} = 23 \cdot 2^{-8k} \cdot y^{p/2} ,$$

where we have used (13.15) and (13.2). Putting $\rho = \sum_{j=1}^{I} a_j e^{i(\lambda_j - \lambda_1)x_o}$ the lemma is proved. □

Lemma 13.8. *If $(n,\omega^*) \notin G_k^*$ and $\omega^* \notin X_k^* \cup Y^*$, then the functions $Q_0^k(x;\omega)$ associated with each of the four subintervals ω of ω^* , for which $4m(\omega) = m(\omega^*)$, are identical.*

<u>**Proof.**</u> Let ω_o and ω be any two of these subintervals. Suppose that $a\,e^{i\lambda x}$ is a term from $Q_0^k(x;\omega_o)$, i.e. there exists an ω', such that $\omega_o \subseteq \omega'$ and $(\psi(\lambda;\omega'),\omega') \in G_k$ and $|n-\psi(\lambda;\omega_o)| < 2^{10k}$, so (13.6) in definition 13.2 of $\tilde{G}_k^1$ is satisfied for (n,ω_o). But since $(n,\omega_o) \notin \tilde{G}_k$ [cf. remark 13.6] condition (13.7) cannot be fulfilled, so

$$m(\omega') > 2^{10k}\, m(\omega) \quad . \tag{13.16}$$

If $\omega \nsubseteq \omega'$ then $\omega \subseteq Y^*$. In fact, $F^k_{\omega'}$ consists of two intervals each of length $2 \cdot 2^{-3k}\, m(\omega') > 2 \cdot 2^{7k}\, m(\omega_o)$ and since $\omega_o \subset \omega'$ and $\omega \nsubseteq \omega'$ and $\omega_o \cup \omega \subset \omega^*$, where $m(\omega^*) = 4m(\omega_o) < 2 \cdot 2^{7k}\, m(\omega_o)$, the smoothing interval ω^* must contain an end point of ω', and we conclude that even $\omega^* \subseteq F^k_{\omega'} \subseteq Y^*$, which is contradicting our assumption. Hence $\omega' \supset \omega^*$ and especially $\omega' \supset \omega$, and the term $a\,e^{i\lambda x}$ must also belong to $P_k(x;\omega)$. As ω_o and ω were arbitrarily chosen, the lemma is proved. □

It follows from the two last lemmata that the approximating term $\rho\, e^{i\lambda x}$ in lemma 13.7 may be chosen with the same ρ and λ for all four subintervals.

<u>§ 14. Construction of the splitting $\Omega(p^*,r)$ of ω^*.</u>

If $n \in \mathbb{N}_o$ we shall write $p^* = (\psi^*(n;\omega^*),\omega^*)$ for short in the following. Let $r \in \mathbb{N}$ be fixed. Let $L = L(p) \in \mathbb{N}$ be the constant defined in (11.6), and let G_r^* be the index set defined by (13.13). Since $L \cdot r \in \mathbb{N}$ we may define

$$\tilde{G}(r) = \left\{ p^* = (\psi^*(n;\omega^*),\omega^*) \in G^*_{rL} \;\middle|\; C^*_{\psi^*(n;\omega^*)}(\omega^*;\chi_F^o) < 2 \cdot 2^{-r} y \right\} . \tag{14.1}$$

Using this subset $\tilde{G}(r)$ of the index set G^*_{rL} we shall define our splitting $\Omega(p^*,r)$ of ω^*, i.e. we shall define a disjoint covering of ω^*. If $p^* = (\psi^*(n;\omega^*),\omega^*) \in \tilde{G}(r)$ it follows immediately that

$$C_{\psi(n;\omega)}(\omega;\chi_F^o) < 2 \cdot 2^{-r} y \tag{14.2}$$

for each of the four dyadic subintervals $\omega \subset \omega^*$, for which $4m(\omega) = m(\omega^*)$. Let ω be any of these subintervals. We consider the two dyadic subintervals ω_1' and ω_2' of ω, for which $2m(\omega_i') = m(\omega)$, $i = 1,2$. If ω_1' and ω_2' both satisfy condition (14.2), we shall iterate this process on each of the intervals ω_1' and ω_2'.

If at least one of the intervals ω_1' and ω_2' does not satisfy (14.2) we shall say that ω is an interval of the splitting $\Omega(p^*,r)$.

In this way we continue until either we have got an element ω satisfying (14.2), such that (14.2) is not fulfilled for at least one of the two subintervals ω_1' and ω_2', or until we have reached level N, i.e. $m(\omega) = 2\pi \cdot 2^{-N}$, where N is the constant given in the introduction to chapter IV. For these intervals from level N, which we also shall define as elements of $\Omega(p^*,r)$, we of course have that condition (14.2) is satisfied. Then it is obvious that $\Omega(p^*,r)$ defines a disjoint covering of ω^*.

The intervals $\omega \in \Omega(p^*,r)$ are characterized by the following obvious lemma.

Lemma 14.1. *Let* $\omega \in \Omega(p^*,r)$. *Then the following three conditions are fulfilled:*

(14.3) $m(\omega) \geqq 2\pi \cdot 2^{-N}$.

(14.4) *If* $\frac{1}{4} m(\omega^*) \geqq m(\omega) \geqq 2 \cdot 2\pi \cdot 2^{-N}$ *then one can find at least one dyadic interval* $\omega' \subset \omega$, *such that* $2m(\omega') = m(\omega)$, *and such that (14.2) is not fulfilled, i.e.*

$$C_{\psi(n;\omega')}(\omega';\chi_F^o) \geqq 2 \cdot 2^{-r} \cdot y .$$

(14.5) *If* ω' *is any dyadic interval, such that* $\omega \subseteqq \omega' \subset \omega^*$ *and* $4m(\omega') \leqq m(\omega^*)$, *then* ω' *satisfies (14.2).*

Let x belong to the middle half of ω^*. We consider the class of smoothing intervals $\tilde{\omega}^* = \omega_{j\nu} \cup \omega_{j+1,\nu}$, where either $\omega_{j\nu} \in \Omega(p^*,r)$ or $\omega_{j+1,\nu} \in \Omega(p^*,r)$. Since $\Omega(p^*,r)$ is a disjoint covering of ω^* with dyadic (half-open) intervals, there exists at least one such interval $\tilde{\omega}^*$ with x lying in its middle half. We define $\omega^*(x)$ as that interval $\tilde{\omega}^*$ satisfying the conditions above, for which $m(\tilde{\omega}^*)$ is maximal. This interval $\omega^*(x)$ is called *the central interval with respect to* x *and the splitting* $\Omega(p^*,r)$.

Lemma 14.2. *The central interval* $\omega^*(x)$ *with respect to* x *and the splitting* $\Omega(p^*,r)$ *, where* $p^* = (\psi^*(n;\omega^*),\omega^*)$ *and* x *belongs to the middle half of* ω^* *, satisfies the following conditions:*

(14.6) $2m(\omega^*(x)) \leqq m(\omega^*)$.

(14.7) x *belongs to the middle half of* $\omega^*(x)$;

(14.8) $\omega^*(x)$ *is a union of intervals from* $\Omega(p^*,r)$;

(14.9) *If* $\omega \in \Omega(p^*,r)$ *and* $\omega \subseteq \omega^*\backslash\omega^*(x)$ *, then* $dist(x;\omega) \geqq \frac{1}{2} m(\omega)$.

Proof. As the biggest interval from $\Omega(p^*,r)$ has measure $\leqq \frac{1}{4} m(\omega^*)$ and $\omega^*(x)$ is the union of one interval from $\Omega(p^*,r)$ and a neighbouring dyadic interval, condition (14.6) is trivial. Furthermore, (14.7) follows from the definition.

Suppose $\omega^*(x) = \omega_{j\nu} \cup \omega_{j+1,\nu}$, where $\omega_{j\nu} \in \Omega^*(p^*,r)$, say , and suppose that $\omega^*(x)$ is not a union of intervals from $\Omega(p^*,r)$. Then $\omega_{j+1,\nu}$ must be contained in an interval $\omega' \in \Omega(p^*,r)$, because of our dyadic partition. Let ω'' be the neighbouring dyadic interval for ω' , for which $\omega_{j\nu} \subseteq \omega''$. Then $\omega^*(x)$ would lie in the middle of $\omega' \cup \omega'' = \tilde{\omega}^*$, thus x would lie in the middle half of $\tilde{\omega}^*$, which is a contradiction to the maximality of $\omega^*(x)$. Hence we have proved (14.8).

Finally, suppose that $\omega \in \Omega(p^*,r)$ and $\omega \subseteq \omega^*\backslash\omega^*(x)$, while $dist(x,\omega) < \frac{1}{2} m(\omega)$. Let ω' be the neighbouring dyadic interval for ω , for which $x \in \omega'$. Then x would lie in the middle half of $\omega' \cup \omega = \tilde{\omega}^*$, and due to the maximality of $\omega^*(x)$ we conclude that $\omega \subset \tilde{\omega}^* \subseteq \omega^*(x)$, contradicting the assumption, and the lemma is proved. □

Lemma 14.3. *Let* $\omega^*(x)$ *be the central interval with respect to* x *and the splitting* $\Omega(p^*,r)$.

If $\omega^*(x) = \omega_{j\nu} \cup \omega_{j+1,\nu}$ *, where at least one of the intervals* $\omega_{j\nu}$ *and* $\omega_{j+1,\nu}$ *belongs to* $\Omega(p^*,r)$ *then*

$$(14.10) \quad \max\left\{C_{\psi(n;\omega_{j\nu})}(\omega_{j\nu};\chi_F^o)\,,\,C_{\psi(n;\omega_{j+1,\nu})}(\omega_{j+1,\nu};\chi_F^o)\right\} < 2\cdot 2^{-r}y\ .$$

If $m(\omega^*(x)) > 2\cdot 2\pi\cdot 2^{-N}$, *then*

$$(14.11) \qquad C^*_{\psi^*(n;\omega^*(x))}(\omega^*(x);\chi_F^o) \geqq 2\cdot 2^{-r}y\ .$$

<u>Proof.</u> We may assume that $\omega_{j\nu}\in\Omega(p^*,r)$. Then by construction

$$C_{\psi(n;\omega_{j\nu})}(\omega_{j\nu};\chi_F^o) < 2\cdot 2^{-r}y\ .$$

By (14.8), the other interval $\omega_{j+1,\nu}$ is a union of intervals from $\Omega(p^*,r)$. Let $\omega'\in\Omega(p^*,r)$ satisfy $\omega'\subseteqq\omega_{j+1,\nu}$ $[\subseteqq\omega^*$, and $4m(\omega_{j+1,\omega})\leqq m(\omega^*)]$. Then by (14.5)

$$C_{\psi(n;\omega_{j+1,\nu})}(\omega_{j+1,\nu};\chi_F^o) < 2\cdot 2^{-r}y\ ,$$

and (14.10) follows.

Finally, if $m(\omega^*(x)) > 2\cdot 2\pi\cdot 2^{-N}$ and $\omega^*(x)=\omega_1\cup\omega_2$, then ω_1 and ω_2 do not belong to level N . We may assume that $\omega_1\in\Omega(p^*,r)$. Using (14.4) we infer that there exists a dyadic interval $\omega'\subset\omega_1$, such that $2m(\omega')=m(\omega_1)$, and such that

$$C_{\psi(n;\omega')}(\omega';\chi_F^o) \geqq 2\cdot 2^{-r}y\ .$$

As ω' is one of the four dyadic subintervals of $\omega^*(x)$ for which $4m(\omega')=m(\omega^*(x))$, we infer that

$$\begin{aligned} C^*_{\psi^*(n;\omega^*(x))}(\omega^*(x);\chi_F^o) &= \max\left\{C_{\psi(n;\omega'')}(\omega'';\chi_F^o)\ \middle|\ \omega''\subset\omega^*(x)\ ,\ 4m(\omega'')=m(\omega^*(x))\right\} \\ &\geqq 2\cdot 2^{-r}y\ , \end{aligned}$$

and we have proved (14.11) and thus the lemma. □

§ 15. Construction of the sets T^* and U^* and E_N.

In this section we shall construct the exceptional set E_N and prove the existence of a constant $C_p \in \mathbb{R}_+$, independent of N, y and F, such that

$$m(E_N) \leqq C_p^p \cdot y^{-p} \cdot m(F) .$$

The set E_N is the union of the previously defined sets S^*, V_L^*, X^* and Y^* together with two other sets T^* and U^*, which we shall define in the sequel.

Let $r \in \mathbb{N}$. Suppose that $p^* = (\psi^*(n;\omega^*),\omega^*) \in \tilde{G}(r)$, where $n \in \mathbb{N}_0$ and ω^* is a smoothing interval held fixed in the following. Let $\Omega(p^*,r)$ be the splitting of ω^* with respect to n and $\overset{o}{\chi}_F$ defined in § 14.

If $\omega_j \in \Omega(p^*,r)$, let t_j denote the midpoint of ω_j and let $\delta_j = m(\omega_j)$, i.e. $\delta_j \geqq 2\pi \cdot 2^{-N}$. For each $x \in \omega^*$ we define a subset $\Omega(x)$ of $\Omega(p^*,r)$ by

$$\Omega(x) = \left\{\omega_j \in \Omega(p^*,r) \mid \forall\, t \in \omega_j :\ |x-t| \geqq \tfrac{1}{2}\delta_j\right\} . \tag{15.1}$$

Using this subset $\Omega(x)$ of disjoint intervals we define a function $\Delta(x)$ for $x \in \omega^*$ by

$$\Delta(x) = \Delta(x;\Omega) \doteq \sum_{\omega_j \in \Omega(x)} \frac{\delta_j^2}{(x-t_j)^2+\delta_j^2} . \tag{15.2}$$

Lemma 15.1. *We have the following estimate for all* $\lambda \in \mathbb{R}_+$,

$$m(\{x \in \omega^* \mid \Delta(x) \geq \lambda\}) \leqq \frac{40}{\lambda}\, exp(-\frac{\lambda}{40}) \cdot m(\omega^*) .$$

Proof. As t_j is the midpoint of ω_j and $dist(t_j;C\omega_j) = \frac{1}{2}\delta_j$ it follows from (15.1) for all $\omega_j \in \Omega(x)$ that

$$|x-t_j| \geqq \delta_j \quad \text{for } \omega_j \in \Omega(x) , \tag{15.3}$$

from which we derive that

$$|x-t| \leq |x-t_j| + |t_j - t| \leq 2|x-t_j| \quad \text{for} \quad t \in \omega_j \; . \tag{15.4}$$

If

$$g_\Omega(x) = \sum_{\omega_j \in \Omega(x)} \int_{\omega_j} \frac{\delta_j}{(x-t)^2+\delta_j^2}\,dt \; ,$$

we get, using (15.4)

$$g_\Omega(x) \geq \sum_{\omega_j \in \Omega(x)} \int_{\omega_j} \frac{\delta_j}{4(x-t_j)^2+\delta_j^2}\,dt \geq \frac{1}{4} \sum_{\omega_j \in \Omega(x)} \frac{\delta_j^2}{(x-t_j)^2+\delta_j^2} = \frac{1}{4}\,\Delta(x) \; ,$$

so we have

$$\{x \in \omega^* \mid \Delta(x) \geq \lambda\} \subseteq \{x \in \omega^* \mid g_\Omega(x) \geq \tfrac{\lambda}{4}\} = \Gamma_\lambda \, . \tag{15.5}$$

Let φ be any non-negative function, for which $\operatorname{supp}(\varphi) \subseteq \omega^*$ and $\int \varphi(x) \cdot \log^+ \varphi(x)\,dx < +\infty$. Then by (5.4), (5.15) and theorem 2.3

$$\int_{\omega^*} \varphi(x) g_\Omega(x)\,dx = \sum_{\omega_j \in \Omega(x)} \int_{\omega_j} \int_{\omega^*} \frac{\varphi(x)\delta_j}{(x-t)^2+\delta_j^2}\,dx\,dt = \sum_{\omega_j \in \Omega(x)} \int_{\omega_j} (P_{\delta_j}\varphi)(t)\,dt \tag{15.6}$$

$$\leq \sum_{\omega_j \in \Omega(x)} \int_{\omega_j} (\theta\varphi)(t)\,dt \leq \int_{\omega^*} (\theta\varphi)(t)\,dt$$

$$\leq 2m(\omega^*) + 8 \int_{\omega^*} \varphi(x) \log^+ \varphi(x)\,dx \; .$$

If we choose $\varphi(x) = \exp(\frac{\lambda}{40}) \cdot \chi_{\Gamma_\lambda}(x)$ it follows from (15.6) that

$$\exp\left(\frac{\lambda}{40}\right) \cdot \frac{\lambda}{4} \cdot m\left(\Gamma_\lambda\right) \leq \int_{\omega^*} \varphi(x) g_\Omega(x)\,dx \leq 2m(\omega^*) + 8 \cdot \frac{\lambda}{40} \cdot \exp\left(\frac{\lambda}{40}\right) \cdot m\left(\Gamma_\lambda\right) \; ,$$

from which we get by a rearrangement, using (15.5) ,

$$m(\{x \in \omega^* \mid \Delta(x) \geq \lambda\}) \leq m(\Gamma_\lambda) \leq \frac{40}{\lambda} \cdot \exp\left(-\frac{\lambda}{40}\right) \cdot m(\omega^*) \; . \qquad \square$$

Using that $\Omega(p^*,r)$ is a disjoint covering of ω^* we define

$$(15.7) \qquad f_n(t) = \frac{1}{m(\omega(t))} \int_{\omega(t)} \overset{o}{\chi}_F(y)\, e^{-iny}\, dy \,, \qquad t \in \omega^* \,,$$

where $\omega(t) \in \Omega(p^*,r)$ is uniquely determined by the condition $t \in \omega(t)$. It is obvious that $\|f_n\|_\infty \leq 1$. We shall in lemma 15.2 prove another estimate of $\|f_n\|_\infty$, which will be needed in the following. We note that since the intervals $\omega \in \Omega(p^*,r)$ are disjoint and satisfy the inequality $m(\omega) \geq 2\pi \cdot 2^{-N}$, we have at most 2^N elements in $\Omega(p^*,r)$, so $f_n(t)$ is a step function subordinated the splitting $\Omega(p^*,r)$ of ω^* .

<u>Lemma 15.2.</u> *We have the following estimate for* $f_n(t)$,

$$|f_n(t)| \leq c_6 \cdot y \cdot 2^{-r} \qquad \text{for } t \in \omega^*$$

i.e. $\|f_n\|_\infty \leq c_6 \cdot y \cdot 2^{-r}$. *Here* $c_6 = 2c_3$, *where* c_3 *is the constant introduced in lemma 9.5.*

<u>Proof.</u> Let $t \in \omega_{j\nu}$ where $\omega_{j\nu} = \omega(t) \in \Omega(p^*,r)$ belongs to level ν . Using the definition (9.3) of the generalized Fourier coefficients, lemma 9.5 and condition (14.2), which is satisfied for intervals from $\Omega(p^*,r)$, we get

$$|f_n(t)| = |c_{n\cdot 2^{-\nu}}(\omega_{j\nu})| \leq c_3 \cdot C_{\psi(n;\Omega)}(\omega; \overset{o}{\chi}_F) < 2c_3 \cdot 2^{-r} \cdot y \,. \qquad \square$$

<u>Remark 15.3.</u> A careful analysis which uses remark 9.3 and the proofs of lemma 9.4 and lemma 9.5 shows that we may choose $c_6 = 2880$.

Let

$$(15.8) \qquad C = 20 \cdot \log 2 \cdot \max\left\{40 \,, \frac{c_3}{c_2}\right\} ,$$

where c_2 is the positive constant occurring in the exponent in corollary 8.5 and c_3 is the constant from lemma 9.5. Using the function $\Delta(x)$

given by (15.2) we define

$$(15.9) \qquad U^*(p^*) = \{x \in \omega^* \mid \Delta(x) > CLr\} ,$$

where $L = L(p)$ is the constant introduced in (11.6). We note that $\Delta(x)$ depends on the choice of p^* .

Let

$$(15.10) \qquad \hat{h}_n(x) = \hat{H}_{\omega^*} f_n(x) = \sup_{\sigma_x} \left| \frac{1}{\pi} (pv) \int_{\sigma_x} \frac{f_n^o(t)}{x-t} dt \right| ,$$

where $\hat{H}_{\omega^*}$ is the modified Hilbert transform with respect to ω^* introduced in (8.9), i.e. the supremum is taken over all smoothing intervals $\sigma_x \subseteqq \omega^*$ for which x belongs to the middle half of σ_x . We define

$$(15.11) \qquad T^*(p^*) = \{x \in \omega^* \mid \hat{h}_n(x) > 2\, CLr \cdot 2^{-r} y\}$$

Let $c_7 = \max \left\{\frac{1}{20 \log 2} , c_1\right\}$, where the constant c_1 was introduced in corollary 8.5.

<u>Lemma 15.4.</u> *The sets $T^*(p^*)$ and $U^*(p^*)$ satisfy the estimates*

$$m(T^*(p^*)) \leqq c_7 \cdot 2^{-20Lr} \cdot m(\omega^*) , \qquad m(U^*(p^*)) \leqq c_7 \cdot 2^{-20Lr} \cdot m(\omega^*) .$$

<u>Proof.</u> We first consider the set $T^*(p^*)$. From lemma 15.2 follows that

$$- c_2 \cdot \frac{y \cdot 2^{-r}}{\|f_n\|_\infty} \cdot 2\, CLr \leqq - \frac{c_2}{2c_3} \cdot 2\, CLr ,$$

hence according to corollary 8.5 and the definition (15.8) of C ,

$$m(T^*(p^*)) \leqq c_1 \cdot m(\omega^*) \cdot \exp\left(- c_2 \cdot \frac{1}{\|f_n\|_\infty} \cdot 2\, CLr\, 2^{-r} y\right)$$

$$\leqq c_1 \cdot m(\omega^*) \cdot \exp\left(- \frac{c_2}{c_3} CLr\right) \leqq c_7 \cdot 2^{-20Lr} m(\omega^*) .$$

Next, using lemma 15.1 we get for $U^*(p^*)$

$$m(U^*(p^*)) \leqq \frac{40}{CLr}\exp\left(-\frac{CLr}{40}\right)\cdot m(\omega^*) \leqq c_7 \cdot 2^{-20Lr} \cdot m(\omega^*) \ . \qquad \square$$

The sets T^* and U^* are now defined by

$$T^* = \bigcup_{r=1}^{+\infty} \left\{ \bigcup_{p^* \in \tilde{G}(r)} T^*(p^*) \right\} , \qquad U^* = \bigcup_{r=1}^{+\infty} \left\{ \bigcup_{p^* \in \tilde{G}(r)} U^*(p^*) \right\} .$$

Theorem 15.5. *$m(T^* \cup U^*) \leqq c_8 \cdot y^{-p}\, m(F)$, where $c_8 = 1000 \cdot c_7$.*

Proof. By the definition (14.1) of $\tilde{G}(r)$ we have $\tilde{G}(r) \subseteqq G^*_{rL}$ for every $r \in \mathbb{N}$. Hence it follows from lemma 13.5 that if we put $p^* = (\psi^*(n;\omega^*),\omega^*)$ for short, then

$$\sum_{p^* \in \tilde{G}(r)} m(\omega^*) \leqq \sum_{p^* \in G^*_{rL}} m(\omega^*) \leqq 500 \cdot 2^{19rL}\, y^{-p}\, m(F) \ .$$

From lemma 15.4 follows that

$$\begin{aligned} m\left(\bigcup_{p^* \in \tilde{G}(r)} \{T^*(p^*) \cup U^*(p^*)\}\right) &\leqq 2c_7 \cdot 2^{-20Lr} \cdot \sum_{p^* \in \tilde{G}(r)} m(\omega^*) \\ &\leqq 2 \cdot c_7 \cdot 2^{-20Lr} \cdot 500 \cdot 2^{19Lr}\, y^{-p}\, m(F) \\ &= 1000 \cdot c_7 \cdot 2^{-Lr} \cdot y^{-p}\, m(F) \ , \end{aligned}$$

so

$$m(T^* \cup U^*) \leqq 1000 \cdot c_7 \cdot y^{-p}\, m(F) \sum_{r=1}^{+\infty} 2^{-rL} \leqq 1000 \cdot c_7 \cdot y^{-p} \cdot m(F) \ . \qquad \square$$

Finally we define the exceptional set E_N by

(15.12) $$E_N = S^* \cup T^* \cup U^* \cup V^*_L \cup X^* \cup Y^* \ .$$

Theorem 15.6. *There exists a constant* $C_p \in \mathbb{R}_+$ *, such that*

$$m(E_N) \leq C_p^p\, y^{-p}\, m(F) .$$

Proof. The theorem follows immediately from theorem 11.1, theorem 11.5, theorem 12.5, theorem 12.6 and theorem 15.5. □

For $x \notin T^*(p^*) \cup U^*(p^*)$ we shall need an estimate of $|S_n^*(x;\chi_F^o;\omega_o^*)|$ compared with $|S_n^*(x;\chi_F^o;\omega^*(x))|$, where $\omega^*(x) \subseteq \omega_o^* \subseteq \omega^*$. More precisely we have

Theorem 15.7. *Let* $x \in\,]-\pi, \pi]$ *and* $x \notin T^*(p^*) \cup U^*(p^*)$ *, let* $\omega^*(x)$ *be the central interval with respect to* x *and the splitting* $\Omega(p^*,r)$ *, and let* ω_o^* *be any smoothing interval satisfying the following conditions:*

i) x *belongs to the middle half of* ω_o^* *;*

ii) $\omega^*(x) \subseteq \omega_o^* \subseteq \omega^*$ *;*

iii) $\omega_o^* \setminus \omega^*(x)$ *is a union of intervals from* $\Omega(p^*,r)$ *.*

Then there exists a constant $c_9 \in \mathbb{R}_+$ *, such that*

$$(15.13) \qquad \Big|\, |S_n^*(x;\chi_F^o;\omega_o^*)| - |S_n^*(x;\chi_F^o;\omega^*(x))| \,\Big| \leq c_9 \cdot Lr \cdot 2^{-r} y .$$

Proof. The estimate (15.13) is trivial if $\omega_o^* = \omega^*(x)$, so we may in the following assume that $\omega^*(x) \subset \omega_o^*$ and $\omega^*(x) \neq \omega_o^*$. Now

$$(15.14) \qquad \Big|\, |S_n^*(x;\chi_F^o;\omega_o^*)| - |S_n^*(x;\chi_F^o;\omega^*(x))| \,\Big| \leq |S_n^*(x;\chi_F^o;\omega_o^*) - S_n^*(x;\chi_F^o;\omega^*(x))|$$

$$= \frac{1}{\pi} \left| \int_{\omega_o^* \setminus \omega^*(x)} \frac{e^{-int}\chi_F^o(t)}{x-t}\, dt \right|$$

[cf. (10.3)], so we shall only estimate

$$\int_{\omega_o^* \setminus \omega^*(x)} \frac{e^{-int}\chi_F^o(t)}{x-t}\, dt = h_n(x) + r_n(x) \quad ,$$

where

$$h_n(x) = \int_{\omega_o^*\setminus\omega^*(x)} \frac{f_n(t)}{x-t}\,dt = (pv)\int_{\omega_o^*} \frac{f_n(t)}{x-t}\,dt - (pv)\int_{\omega^*(x)} \frac{f_n(t)}{x-t}\,dt$$

and

$$(15.15) \qquad r_n(x) = \int_{\omega_o^*\setminus\omega^*(x)} \frac{e^{-int}\chi_F^o(t) - f_n(t)}{x-t}\,dt \ .$$

Here we have used $f_n(t)$ defined in (15.7). As ω_o^* and $\omega^*(x)$ are intervals of σ_x-type [cf. (15.10)] we immediately get, since $x \notin T^*(p^*)$,

$$(15.16) \qquad |h_n(x)| \leqq 2\hat{h}_n(x) \leqq 4 \cdot CLr \cdot 2^{-r} y \ .$$

Let us turn to $r_n(x)$ defined in (15.15). As above we let $\delta_j = m(\omega_j)$ for $\omega_j \in \Omega(p^*,r)$ and let t_j denote the midpoint of ω_j . By assumption $\omega_o^*\setminus\omega^*(x)$ is a union of intervals from $\Omega(p^*,r)$,

$$\omega_o^*\setminus\omega^*(x) = \bigcup_{\omega_j \in \Omega(x)} \omega_j \ , \qquad \Omega(x) \subset \Omega(p^*,r) \ .$$

Using (14.9) in lemma 14.2 we get $\text{dist}(x;\omega_j) \geqq \frac{1}{2}\delta_j$ for each $\omega_j \in \Omega(x)$, so the index set $\Omega(x)$ above is identical with the index set introduced in (15.1). Especially, it follows from lemma 15.2 that

$$(15.17) \qquad |f_n(t)| \leqq c_6 \cdot 2^{-r} y \qquad \text{for all} \quad t \in \omega^* \ .$$

Next we note that

$$\frac{1}{x-t} = \frac{1}{x-t_j} + \frac{t-t_j}{(x-t)(x-t_j)} \ ,$$

and as

$$\int_{\omega_j} \left\{ e^{-int}\chi_F^o(t) - f_n(t) \right\} dt = \int_{\omega_j} e^{-int}\chi_F^o(t)\,dt - \frac{1}{m(\omega_j)} \int_{\omega_j} \chi_F^o(y) e^{-iny}\,dy \cdot m(\omega_j) = 0$$

we infer that

$$\int_{\omega_j} \frac{e^{-int}\chi_F^o(t) - f_n(t)}{x - t_j}\,dt = 0 .$$

Hence

$$(15.18) \qquad r_n(x) = \int_{\omega_o^*\setminus\omega^*(x)} \frac{e^{-int}\chi_F^o(t) - f_n(t)}{x - t}\,dt$$

$$= \sum_{\omega_j \in \Omega(x)} \int_{\omega_j} \left\{ \frac{1}{x-t_j} + \frac{t-t_j}{(x-t)(x-t_j)} \right\} \left\{ e^{-int}\chi_F^o(t) - f_n(t) \right\} dt$$

$$= \sum_{\omega_j \in \Omega(x)} \int_{\omega_j} \frac{t-t_j}{(x-t)(x-t_j)} e^{-int}\chi_F^o(t)\,dt$$

$$- \sum_{\omega_j \in \Omega(x)} \int_{\omega_j} \frac{t-t_j}{(x-t)(x-t_j)} f_n(t)\,dt .$$

Since $|x - t| \geqq \frac{1}{2}\delta_j$ for all $t \in \omega_j$, we get $|x - t| \geqq |x - t_j| - \frac{1}{2}\delta_j$ and so by (15.3)

$$(15.19) \qquad |(x-t)(x-t_j)| \geqq (x-t_j)^2 - \frac{1}{2}\delta_j|x-t_j|$$

$$\geqq (x-t_j)^2 - \frac{1}{2}(x-t_j)^2 = \frac{1}{2}(x-t_j)^2 \geqq \frac{1}{4}\{(x-t_j)^2 + \delta_j^2\} ,$$

from which we derive for $t \in \omega_j$,

$$(15.20) \qquad \left| \frac{t-t_j}{(x-t)(x-t_j)} \right| \leqq \frac{1}{2}\delta_j \cdot \frac{1}{\frac{1}{4}\{(x-t_j)^2 + \delta_j^2\}} = \frac{2\delta_j}{(x-t_j)^2 + \delta_j^2} .$$

Using this inequality together with (15.17) we get for the last term in (15.18) as $x \notin U^*(p^*)$,

$$(15.21) \qquad \left| \sum_{\omega_j \in \Omega(x)} \int_{\omega_j} \frac{t-t_j}{(x-t)(x-t_j)} f_n(t)\,dt \right| \leqq c_6 \cdot 2^{-r} y \sum_{\omega_j \in \Omega(x)} \frac{2\delta_j^2}{(x-t_j)^2 + \delta_j^2}$$

$$\leqq 2c_6 \cdot 2^{-r} y\, \Delta(x) \leqq 2c_6 \cdot CLr \cdot 2^{-r} y .$$

Finally, we shall prove an estimate of the same type for the first term in (15.18), i.e. for

$$(15.22)\qquad \sum_{\omega_j \in \Omega(x)} \int_{\omega_j} \frac{t-t_j}{(x-t)(x-t_j)} e^{-int} \chi_F^0(t)dt \ .$$

Consider the function

$$(15.23)\quad \varphi_j(t) = \frac{t-t_j}{(x-t)(x-t_j)} \exp\left(-i\{n - \frac{2\pi}{m(\omega_j)} \psi(n;\omega_j)\} t\right), \qquad t \in \omega_j \ ,$$

where $\omega_j \in \Omega(x)$. For convenience, let $\delta_j = m(\omega_j) = 2\pi \cdot 2^{-\nu}$, i.e. $\frac{2\pi}{m(\omega_j)} = 2^{\nu}$. As $x \notin \bar{\omega}_j$, it is obvious that $\varphi \in C^2(\bar{\omega}_j)$, so using lemma 9.4 we infer that

$$\varphi_j(t) = \sum_{\mu \in Z} \bar{\gamma}_{j\mu} \exp\left(-i \cdot 2^{\nu} \cdot \frac{\mu}{3} \cdot t\right), \qquad t \in \omega_j \ ,$$

where

$$(15.24)\qquad (1+\mu^2)|\gamma_{j\mu}| \leqq c_2 \cdot \{\max_{\omega_j} |\varphi_j| + 2^{-2\nu} \max_{\omega_j} |\varphi_j''|\} \ .$$

From (15.23) and (15.20) follows

$$(15.25)\qquad |\varphi_j(t)| \leqq \frac{2\delta_j}{(x-t_j)^2 + \delta_j^2} \ ,$$

and using that $|n - 2^{\nu} \psi(n;\omega_j)| \leqq 2^{\nu}$ we get after a small computation

$$|\varphi_j''(t)| \leqq 2^{2\nu} |\varphi_j(t)| + 2 \cdot 2^{\nu} \cdot \frac{1}{|x-t_j||x-t|} + 2 \cdot 2^{\nu} \left|\frac{t-t_j}{(x-t)^2(x-t_j)}\right|$$
$$+ 2\left|\frac{1}{(x-t)^2(x-t_j)}\right| + 2\left|\frac{t-t_j}{(x-t)^3(x-t_j)}\right| \ ,$$

so using (15.19), (15.20), (15.25) and that $|x-t| \geqq \frac{1}{2}\delta_j = \pi \cdot 2^{-\nu}$ we get

$$2^{-2\nu}|\varphi_j''(t)| \leqq \left(2 + \frac{4}{\pi} + \frac{4}{\pi} + \frac{4}{\pi^2} + \frac{4}{\pi^2}\right) \cdot \frac{\delta_j}{(x-t_j)^2 + \delta_j^2} \leqq 6 \cdot \frac{\delta_j}{(x-t_j)^2 + \delta_j^2} \ .$$

Substituting this inequality and (15.25) into (15.24) we get

$$(1+\mu^2)|\gamma_{j\mu}| \leq c_{10} \cdot \frac{\delta_j}{(x-t_j)^2+\delta_j^2} \quad .$$

Using this inequality we get the following estimate of (15.22)

$$\left| \sum_{\omega_j \in \Omega(x)} \int_{\omega_j} \frac{t-t_j}{(x-t)(x-t_j)} e^{-int} \chi_F^o(t)dt \right|$$

$$= \left| \sum_{\omega_j \in \Omega(x)} \int_{\omega_j} \frac{t-t_j}{(x-t)(x-t_j)} \exp\left(-i\{n-2^\nu \psi(n;\omega_j)\}t\right) \cdot \exp\left(-i\, 2^\nu \psi(n;\omega_j)t\right) \cdot \chi_F^o(t)dt \right|$$

$$= \left| \sum_{\omega_j \in \Omega(x)} \int_{\omega_j} \varphi_j(t) \cdot \exp\left(-i\, 2^\nu \psi(n;\omega_j)t\right) \cdot \chi_F^o(t)dt \right|$$

$$= \left| \sum_{\omega_j \in \Omega(x)} \int_{\omega_j} \sum_{\mu \in Z} \bar{\gamma}_{j\mu} \exp\left(-i\, 2^\nu \{\psi(n;\omega_j) + \tfrac{\mu}{3}\}t\right) \cdot \chi_F^o(t)dt \right|$$

$$= \left| \sum_{\omega_j \in \Omega(x)} \sum_{\mu \in Z} \bar{\gamma}_{j\mu}\, m(\omega_j) \cdot c_{\psi(n;\omega_j)+\frac{\mu}{3}}(\omega_j;\chi_F^o) \right|$$

$$= \left| \sum_{\omega_j \in \Omega(x)} \sum_{\mu \in Z} (1+\mu^2)\bar{\gamma}_{j\mu}\, \delta_j \cdot \frac{1}{1+\mu^2} \cdot c_{\psi(n;\omega_j)+\frac{\mu}{3}}(\omega_j;\chi_F^o) \right|$$

$$\leq c_{10} \cdot \sum_{\omega_j \in \Omega(x)} \left\{ \frac{\delta_j^2}{(x-t_j)^2+\delta_j^2} \cdot \sum_{\mu \in Z} \frac{1}{1+\mu^2} \left| c_{\psi(n;\omega_j)+\frac{\mu}{3}}(\omega_j;\chi_F^o) \right| \right\}$$

$$\leq 10 \cdot c_{10} \cdot \sum_{\omega_j \in \Omega(x)} \frac{\delta_j^2}{(x-t_j)^2+\delta_j^2} \cdot C_{\psi(n;\omega_j)}(\omega_j;\chi_F^o)$$

$$\leq 10 \cdot c_{10} \cdot 2 \cdot 2^{-r} y \cdot \sum_{\omega_j \in \Omega(x)} \frac{\delta_j^2}{(x-t_j)^2+\delta_j^2} = 20 \cdot c_{10} \cdot 2^{-r} y\, \Delta(x) \leq 20 \cdot c_{10} \cdot 2^{-r} \cdot CLr \cdot y \ ,$$

where we have used that $C_{\psi(n;\omega_j)}(\omega_j;\chi_F^o) < 2 \cdot 2^{-r} y$ by (14.2) and that $x \notin U^*(p^*)$, so $\Delta(x) \leq C \cdot L \cdot r$. Thus, by means of (15.14), (15.16), (15.21) and the estimate above we get

$$\Big| |S_n^*(x;\chi_F^o;\omega_o^*)| - |S_n^*(x;\chi_F^o;\omega^*(x))| \Big| \leq c_9 \cdot CLr \cdot 2^{-r} y \ . \qquad \square$$

Remark 15.8. By using all the previously determined constants we can prove that we may choose $c_9 = 32$.

§ 16. Estimation for elements $p^* \notin G^*_{rL}$.

We have defined the exceptional set E_N in (15.12) and proved an estimate for this set in theorem 15.5. Now, consider a point $x \notin E_N$ and let x belong to the middle half of a smoothing interval ω_0^* , where $m(\omega_0^*) = 2 \cdot 2\pi \cdot 2^{-\nu}$ and $\nu \leq N-1$. Let n_0 be any frequency associated with ω_0^* , i.e. there exists an $m_0 \in \mathbb{N}_0$, such that $n_0 = m_0 \cdot 2^{\nu+1}$, where the extra factor 2 comes from the fact that we are considering a smoothing interval consisting of two dyadic intervals from level ν . This means that $\psi^*(n_0;\omega_0^*) = m_0$. If $(m_0,\omega_0^*) \in G^*_{rL}$ for some r , we can use our functions P_{rL} defined in § 12. This condition is of course not satisfied in general, so in this section we shall prove two theorems, which ensure that if $p_0^* = (m_0;\omega_0^*) \notin G^*_{rL}$ then we can find an element $\tilde{p}^* \in G^*_{rL}$, which in some sense does not lie "too far away" from the given element $p_0^* \notin G^*_{rL}$. It is here essential that we in (11.6) have defined $L \in \mathbb{N}$, such that $L \geq 100$.

Theorem 16.1. *Let $x \notin E_N$ and suppose that x belongs to the middle half of a smoothing interval ω_0^* , where $m(\omega_0^*) = 2 \cdot 2\pi \cdot 2^{-\nu}$, $\nu \leq N-1$. Let $n_0 = m_0 \cdot 2^{\nu+1}$, $m_0 \in \mathbb{N}_0$ and assume that there exists an $r \in \mathbb{N}$, such that $p_0^* = (m_0,\omega_0^*) \notin G^*_{rL}$ and*

$$(16.1) \qquad 2^{-r} y \leq C^*(p^*) = C^*_{m_o}(\omega_o^*;\chi_F^o) < 2 \cdot 2^{-r} y \,.$$

Then there exist a smoothing interval $\tilde{\omega}^* \supseteq \omega_0^*$ containing x in its middle half, $m(\tilde{\omega}^*) = 2 \cdot 2\pi \cdot 2^{-\mu}$, $\mu \leq \nu$, and two integers $\tilde{n}$ and $\tilde{m}$ satisfying $\tilde{m} = \psi^*(\tilde{n};\tilde{\omega}^*)$, i.e. $\tilde{n} = \tilde{m} \cdot 2^{\mu+1}$, such that $\tilde{p}^* = (\tilde{m},\tilde{\omega}^*) \in G^*_{rL}$ and

$$(16.2) \qquad |\psi^*(\tilde{n};\omega_0^*) - \psi^*(n_0;\omega_0^*)| = |[\tilde{m} \cdot 2^{\mu-\nu}] - m_0| \leq 60 \cdot 2^r \,.$$

Furthermore, if $\tilde{p}_o^* = (\psi^*(\tilde{n};\omega_0^*),\, \omega_0^*)$, then for any n satisfying

$$(16.3) \qquad |\psi^*(n;\omega_0^*) - \psi^*(n_0;\omega_0^*)| = |\psi^*(n;\omega_0^*) - m_0| \leq 120 \cdot 2^{2r}$$

we have

$$(16.4)\qquad \Big|\,|S^*_{n_o}(x;\chi^o_F;\omega^*_o)| - |S^*_n(x;\chi^o_F;\omega^*_o)|\,\Big| \leqq 200\cdot\left\{C^*_{\psi^*(\tilde{n};\omega^*_o)}(\omega^*_o;\chi^o_F) + 2\cdot 2^{-r}y\right\},$$

Proof. As $x \notin E_N$, we have $\omega^*_o \notin S^*$, so by lemma 11.3 we get that (16.1) is in fact satisfied for some $r \in \mathbb{N}$.

Let $\omega'_o \subseteq \omega^*_o$, $4m(\omega'_o) = m(\omega^*_o)$, be one of the subintervals of ω^*_o, for which

$$C_{\psi(n_o;\omega'_o)}(\omega'_o;\chi^o_F) = C^*_{m_o}(\omega^*_o;\chi^o_F),$$

and let $\omega' \subset \omega^*_o$ be any other subinterval of ω^*_o, for which $4m(\omega') = m(\omega^*_o)$. If P_o and P denote the functions

$$P_o = P_{rL}(\cdot;\omega'_o) \qquad \text{and} \qquad P = P_{rL}(\cdot;\omega'),$$

introduced in § 12, it follows from lemma 13.8 that $P = P_o$.

It follows from remark 12.1 that

$$(16.5)\qquad |c_m(\omega';\chi^o_F - P)| \leqq 2^{-rL}\,y^{p/2} \quad \text{for all } m.$$

Since $\omega^*_o \notin S^*$ and $2^{-r}y \leqq C^*_m(\omega^*_o;\chi^o_F)$ it follows from lemma 11.4 that $y^{p/2} \leqq 2^{Lr/4}y$.

It is obvious from the construction of P that $\chi^o_F - P \in L^2(\omega')$, so we may use lemma 9.6. From (16.5) follows that $B = 2^{-rL}y^{p/2}$ and that M may be any positive integer. We choose $M = 2^{10rL}$. In order to find A in lemma 9.6 we shall use (11.3) in lemma 11.2 for $p = 2$. We get

$$\left\{\frac{1}{m(\omega')}\int_{\omega'}|\chi^o_F - P|^2dx\right\}^{\frac12} \leqq \left\{\frac{1}{m(\omega')}\int_{\omega'}|\chi^o_F|^2dx\right\}^{\frac12} + \left\{\frac{1}{m(\omega')}\int_{\omega'}|P|^2dx\right\}^{\frac12}$$

$$\leqq y + \left\{\frac{1}{m(\omega')}\int_{\omega'}|P|^2dx\right\}^{\frac12},$$

so using (13.2) in lemma 13.1 [note that $\omega \notin X_k$ as $x \notin E_N$] we get

$$\left\{\frac{1}{m(\omega')}\int_{\omega'}|\chi^o_F - P|^2dx\right\}^{\frac12} \leqq y + 2^{2rL}y^{p/2} = A.$$

An application of lemma 9.6 now gives for all n [cf. (16.5)]

$$(16.6) \quad C_n(\omega';\chi_F^o-P) \leqq 9\cdot\left\{B\log M+\frac{A}{\sqrt{M}}\right\}$$

$$\leqq 9\cdot\left\{2^{-rL}y^{p/2}\cdot 10rL\cdot\log 2+2^{-5rL}y+2^{-5rL}\cdot 2^{2rL}y^{p/2}\right\}$$

$$\leqq 9\cdot\left\{10\log 2\cdot rL\cdot 2^{-rL}+2^{-3rL}\right\}\cdot 2^{rL/4}y+9\cdot 2^{-5rL}y$$

$$\leqq 2^{-rL/2}y\ ,$$

where we in the latter estimate have used that $L \geqq 100$ and $r \in \mathbb{N}$.

Since $C_{\psi(n_o;\omega_o')}(\omega_o';\chi_F^o) = C^*_{m_o}(\omega_o^*;\chi_F^o) \geqq 2^{-r}y$ and $P = P_o$ we get

$$C_{\psi(n_o;\omega_o')}(\omega_o';P_o)+2^{-rL/2}y \geqq C_{\psi(n_o;\omega_o')}(\omega_o';P_o)+C_{\psi(n_o;\omega_o')}(\omega';\chi_F^o-P)$$

$$\geqq C_{\psi(n_o;\omega_o')}(\omega_o';\chi_F^o) \geqq 2^{-r}y\ ,$$

and hence by rearrangement

$$(16.7) \quad C_{\psi(n_o;\omega_o')}(\omega_o';P_o) \geqq (2^{-r}-2^{-rL/2})y\ .$$

We shall now prove the existence of a frequency λ in P_o, such that $|\psi(\lambda;\omega_o')-\psi(n_o;\omega_o')| < 2^{5rL}$. Suppose that this is not the case, i.e. suppose that every λ satisfies $|\psi(\lambda;\omega_o')-\psi(n_o;\omega_o')| \geqq 2^{5rL}$. Using lemma 9.7 and the second inequality in (13.2) we get, applying once more that $y^{p/2} \leqq 2^{Lr/4}y$,

$$C_{\psi(n_o;\omega_o')}(\omega_o';P_o) \leqq 2^{-5rL}\cdot\sum|a_n| \leqq 2^{-5rL}\cdot 2^{2rL}y^{p/2} \leqq 2^{-2rL}y\ ,$$

which is a contradiction to (16.7), so we have proved that for some λ occurring as a frequency in P_o we have

$$|\psi(\lambda;\omega_o')-\psi(n_o;\omega_o')| < 2^{5rL}\ .$$

We choose one such λ and denote it by $\tilde{n}$. From the construction of P_{rL} in § 12 follows that λ must be associated to some interval $\tilde{\omega}' \supseteq \omega_o'$, such that $(\psi(\tilde{n};\tilde{\omega}'),\tilde{\omega}')\in G_{rL}$. Let $\tilde{\omega}^*$ be a smoothing interval, such that $4m(\tilde{\omega}')=m(\tilde{\omega}^*)$ and such that $x\ (\in\tilde{\omega}')$ is contained in the middle half of $\tilde{\omega}^*$. Then

$$\tilde{\omega}^*\supset\omega_o^*\ ,\qquad m(\tilde{\omega}^*)=4\cdot 2\pi\cdot 2^{-\mu}\ ,\qquad \psi^*(\tilde{n};\tilde{\omega}^*)=\tilde{m}\qquad\text{and}\qquad \tilde{n}=\tilde{m}\cdot 2^{\mu}\ ,$$

and $\tilde{p}^*=(\tilde{m},\tilde{\omega}^*)\in G^*_{rL}$. We shall prove (16.2) for this element $\tilde{p}^*=(\tilde{m},\tilde{\omega}^*)$.

Since $p_o^*\notin G^*_{rL}$ we may write

$$P=\rho\, e^{i\tilde{n}x}+Q_o'(x)+Q_1(x)\ ,$$

where $Q_1(x)$ contains all the terms from P for which the corresponding frequencies λ' satisfy

$$|\psi(n_o;\omega_o')-\psi(\lambda';\omega_o')|\geq 2^{10rL}\ ,$$

and where $Q_o'(x)$ by lemma 13.7 satisfies the estimate (since we already have removed the term $\rho\, e^{i\tilde{n}x}$)

$$|Q_o'(x)|\leq 23\cdot 2^{-8rL}y^{p/2}\leq 4\cdot 2^{-7rL}\ ,$$

where we once more have applied that $y^{p/2}\leq 2^{rL/4}y$. Hence

$$C_{\psi(n;\omega')}(\omega';P-\rho\, e^{i\tilde{n}x})\leq 4\cdot 2^{-7rL}y\quad\text{for}\quad |\psi(n;\omega')-\psi(n_o;\omega')|\leq 2^{9rL}\ ,$$

where ω' is any of the four subintervals of ω_o^* . From (16.6) then follows that

$$\text{(16.8)}\quad C_{\psi(n;\omega')}(\omega';\chi_F^o-\rho\, e^{i\tilde{n}x})\leq C_{\psi(n;\omega')}(\omega';\chi_F^o-P)+C_{\psi(n;\omega')}(\omega';P-\rho\, e^{i\tilde{n}x})$$
$$\leq 2^{-rL/2}y+4\cdot 2^{-7rL}y\leq 2\cdot 2^{-rL/2}y$$

for $|\psi(n;\omega')-\psi(n_o;\omega')|\leq 2^{9rL}$, so using lemma 13.8 and (16.7) we get

$$\text{(16.9)}\quad C_{\psi(n_o;\omega_o')}(\omega_o';\rho\, e^{i\tilde{n}x})\geq C_{\psi(n_o;\omega_o')}(\omega_o';P_o)-C_{\psi(n_o;\omega_o')}(\omega_o';P-\rho\, e^{i\tilde{n}x})$$
$$\geq (2^{-r}-2^{-rL/2})y-4\cdot 2^{-7rL}y\geq \frac{1}{2}\cdot 2^{-r}y$$

for $|\psi(n;\omega')-\psi(n_o;\omega')|\leq 2^{9rL}$.

Now,

$$(16.10)\qquad |\rho| \leqq 10 \cdot C_{\psi(\tilde{n};\omega')}(\omega';\rho\, e^{i\tilde{n}x})$$

$$\leqq 10 \cdot \left\{ C_{\psi(\tilde{n};\omega')}(\omega';\chi_F^o - \rho\, e^{i\tilde{n}x}) + C_{\psi(\tilde{n};\omega')}(\omega';\chi_F^o) \right\}$$

$$\leqq 10 \cdot \left\{ 2 \cdot 2^{-rL/2}\, y + C^*_{\psi(\tilde{n};\tilde{\omega})}(\tilde{\omega};\chi_F^o) \right\} \;,$$

where we have used that (16.8) especially is valid for $n = \tilde{n}$.

Using (11.4) in lemma 11.2 we immediately get

$$(16.11)\qquad |\rho| \leqq 10 \cdot \left\{ 2 \cdot 2^{-rL/2}\, y + y \right\} < 30\, y \;.$$

Using (16.9) and lemma 9.7 [note that here $f(x) = \rho\, e^{i\tilde{n}x}$] we get

$$\frac{1}{2} \cdot 2^{-r}\, y \leqq C_{\psi(n_o;\omega_o')}(\omega_o';\rho\, e^{i\tilde{n}x}) \leqq |\rho| \cdot |\psi(n_o;\omega_o') - \psi(\tilde{n};\omega_o')|^{-1} \;,$$

so using (16.11) above we get

$$|\psi(n_o;\omega_o') - \psi(\tilde{n};\omega_o')| \leqq 2 \cdot |\rho| \cdot 2^r \cdot \frac{1}{y} \leqq 60 \cdot 2^r \;,$$

proving (16.2).

Suppose that n satisfies the condition

$$|\psi^*(n;\omega_o^*) - \psi^*(n_o;\omega_o^*)| \leqq 120 \cdot 2^{2r} \;.$$

We have

$$(16.12)\qquad \left| e^{inx}\, S_n^*(x;\omega_o^*;\chi_F^o) - e^{in_o x}\, S_{n_o}^*(x;\omega_o^*;\chi_F^o) \right|$$

$$\leqq \left| e^{inx}\, S_n^*(x;\omega_o^*;\chi_F^o - \rho\, e^{i\tilde{n}x}) - e^{in_o x}\, S_{n_o}^*(x;\omega_o^*;\chi_F^o - \rho\, e^{i\tilde{n}x}) \right|$$

$$+ \left| S_n^*(x;\omega_o^*;\rho\, e^{i\tilde{n}x}) \right| + \left| S_{n_o}^*(x;\omega_o^*;\rho\, e^{i\tilde{n}x}) \right| \;,$$

where the latter two terms are estimated in the following way by lemma 10.9

and (16.10) [note again that $f(x) = \rho e^{i\tilde{n}x}$]

$$\left|S_n^*(x;\omega_o^*;\rho e^{i\tilde{n}x})\right| + \left|S_{n_o}^*(x;\omega_o^*;\rho e^{i\tilde{n}x})\right| \leq 2 \cdot |\rho| \cdot 10$$

$$\leq 200\left\{2 \cdot 2^{-rL/2} y + C^*_{\psi^*(\tilde{n};\tilde{\omega})}(\tilde{\omega};\chi_F^o)\right\}.$$

The first term on the right hand side of (16.12) is estimated by means of lemma 10.7, applied at most $120 \cdot 2^{2r}$ times, using (16.8) each time. Hence,

$$\left|e^{inx} S_n^*(x;\omega_o^*;\chi_F^o - \rho e^{i\tilde{n}x}) - e^{in_o x} S_{n_o}^*(x;\omega_o^*;\chi_F^o - \rho e^{i\tilde{n}x})\right|$$

$$\leq 120 \cdot 2^{2r} \cdot 2 \cdot 2^{-rL/2} \cdot y \cdot c_5 ,$$

so from (16.12)

$$\left|e^{inx} S_n^*(x;\omega_o^*;\chi_F^o) - e^{in_o x} S_{n_o}^*(x;\omega_o^*;\chi_F^o)\right|$$

$$\leq 240 \cdot c_5 \cdot 2^{2r-Lr/2} y + 400 \cdot 2^{-Lr/2} y + 200 \cdot C^*_{\psi^*(\tilde{n};\tilde{\omega})}(\tilde{\omega};\chi_F^o) ,$$

and (16.4) follows, because $L \geq 100$, and c_5 may be chosen equal to 85,000 . □

In the next theorem we improve the results above, such that we shall also get an estimate for $C^*(\bar{p}^*)$, and we shall be able to involve the splitting $\Omega(\bar{p}^*;r)$ for the constructed element $\bar{p}^*$. The assumptions are the same as in theorem 16.1. We shall, however, repeat all these conditions once more in the formulation of the theorem.

<u>Theorem 16.2.</u> *Let* $x \notin E_N$ *and suppose that* x *belongs to the middle half of a smoothing interval* ω_o^* *, where* $m(\omega_o^*) = 2 \cdot 2\pi \cdot 2^{-\nu}$ *,* $\nu \leq N-1$ *. Let* $n_o = m_o \cdot 2^{\nu+1}$ *,* $m_o \in \mathbb{N}_o$ *, and assume that there exists an* $r \in \mathbb{N}$ *, such that* $p_o^* = (m_o;\omega_o^*) \notin G^*_{rL}$ *and*

$$2^{-r} y \leq C^*(p_o^*) = C^*_{m_o}(\omega_o^*;\chi_F^o) < 2 \cdot 2^{-r} y .$$

Then one can find a smoothing interval $\bar{\omega}^* \supseteq \omega_o^*$ *containing* x *in its middle half,* $m(\bar{\omega}^*) = 2 \cdot 2\pi \cdot 2^{-\mu}$ *,* $\mu \leq \nu$ *, and two integers* $\bar{n}$ *and* $\bar{m}$

satisfying $\bar{m} = \psi^*(\bar{n};\bar{\omega}^*)$, *i.e.* $\bar{n} = \bar{m} \cdot 2^{\mu+1}$, *and another integer* $k \in \{1,2, \ldots , r\}$, *such that* $\bar{p}^* = (\bar{m},\bar{\omega}^*) \in G^*_{kL}$.

If $p^*_0 = (\tilde{m},\tilde{\omega}^*_0) \in G^*_{rL}$ *is given by theorem 16.1 , then*

$$C^*(\tilde{p}^*_0) = C^*_{\tilde{m}}(\omega^*_0;\chi^0_F) < 2^{-(k-1)}y . \tag{16.13}$$

Furthermore,

$$C^*(\bar{p}^*) = C^*_{\bar{m}}(\bar{\omega}^*;\chi^0_F) < 2^{-(k-1)}y \tag{16.14}$$

and the splitting $\Omega(\bar{p}^*;k)$ *of* $\bar{\omega}^*$ *is defined. If* $\bar{\omega}^*(x)$ *denotes the central interval with respect to* x *and* $\Omega(\bar{p}^*;k)$, *then* $\bar{\omega}^*(x) \subseteqq \omega^*_0$, *and* $\bar{\omega}^*(x) \neq \omega^*_0$ *if* x *is not an endpoint for one of the two dyadic intervals defining the middle half of* ω^*_0 , *and* $\omega^*_0\backslash\bar{\omega}^*(x)$ *is a union of intervals from* $\Omega(\bar{p}^*;k)$.

<u>Proof.</u> By Σ we denote the set of triples (n,ω^*,ℓ) , $n \in N$, ω^* a smoothing interval and $\ell \in \{1,2, \ldots , r\}$, satisfying the following four conditions:

i) $\omega^* \supseteqq \omega^*_0$ and x belongs to the middle half of ω^* , and if $m(\omega^*) = 4 \cdot 2\pi \cdot 2^{-\tau}$ then there exists an integer $m \in N$, such that $n = m \cdot 2^{\tau}$, i.e. $\psi^*(n;\omega^*) = m$.

ii) $C^*(\tilde{p}^*_0) = C^*_{\tilde{m}}(\omega^*_0;\chi^0_F) < 2^{-(\ell-1)}y$, where $\tilde{p}^*_0$ is defined by theorem 16.1.

iii) $|\psi^*(n;\omega^*_0) - \psi^*(n_0;\omega^*_0)| \leqq 60 \sum_{j=1}^{r} 2^j < 120 \cdot 2^r$.

iv) $(\psi^*(n;\omega^*), \omega^*) \in G^*_{\ell L}$.

First we claim that $\Sigma \neq \emptyset$. Let $\tilde{n}$ and $\tilde{\omega}^*$ be given by theorem 16.1. If $C^*(\tilde{p}^*_0) = C^*_{\tilde{m}}(\omega^*_0;\chi^0_F) < 2^{-(r-1)}y$, then it follows at once that $(\tilde{n},\tilde{\omega}^*,r) \in \Sigma$.

Suppose that $C^*(\tilde{p}^*_0) \geqq 2^{-(r-1)}y$. From lemma 11.3 follows that one can find $\ell \in \{1,2, \ldots , r-1\}$, such that

$$2^{-\ell}y \leqq C^*(\tilde{p}^*_0) < 2^{-(\ell-1)}y .$$

If $\tilde{p}^*_0 \in G^*_{\ell L}$ we define $\tilde{n}' = \frac{4 \cdot 2\pi}{m(\omega^*_0)} \cdot \psi^*(\tilde{n};\omega^*_0)$, and it is easy to check that $(\tilde{n}', \omega_o^*, \ell) \in \Sigma$.

We are therefore left with the possibility that $\tilde{p}_o^* \notin G_{\ell L}^*$. An application of theorem 16.1 with n_o and r replaced by $\tilde{n}'$ and ℓ gives a new couple $(\tilde{n}_1, \tilde{\omega}_1^*)$, and we claim that the triple $(\tilde{n}_1, \tilde{\omega}_1^*, \ell)$ belongs to Σ . As the conditions i), ii) and iv) are trivial, we only have to check iii). If we insert $\psi^*(\tilde{n}';\omega_o^*)$ and then use the triangle inequality, we get

$$|\psi^*(\tilde{n}_1;\omega_o^*) - \psi^*(n_o;\omega_o^*)| \leq |\psi^*(\tilde{n}_1;\omega_o^*) - \psi^*(\tilde{n}';\omega_o^*)| + |\psi^*(\tilde{n}';\omega_o^*) - \psi^*(n_o;\omega_o^*)|$$

$$\leq |\psi^*(\tilde{n}_1;\tilde{\omega}_1^*) - \psi^*(\tilde{n}';\tilde{\omega}_1^*)| + |\psi^*(\tilde{n}';\omega_o^*) - \psi^*(n_o;\omega_o^*)| \quad ,$$

where we in the latter estimate have used that $\tilde{\omega}_1^* \supset \omega_o^*$, so

$$|\psi^*(n_1;\omega_o^*) - \psi^*(n';\omega_o^*)| = \left|\left[\frac{1}{8\pi} \cdot \tilde{n}_1 \cdot m(\omega_o^*)\right] - \left[\frac{1}{8\pi} \cdot \tilde{n}_1' \cdot m(\omega_o^*)\right]\right|$$

$$\leq \left|\left[\frac{1}{8\pi} \cdot \tilde{n}_1 \cdot m(\tilde{\omega}_1^*)\right] - \left[\frac{1}{8\pi} \cdot \tilde{n}' \cdot m(\omega_o^*)\right]\right|$$

$$= |\psi^*(\tilde{n}_1;\tilde{\omega}_1^*) - \psi^*(\tilde{n}';\tilde{\omega}_1^*)| \quad ,$$

since $\dfrac{m(\omega_1^*)}{m(\omega_o^*)} = 2^s$ for some $s \in \mathbb{N}$, and the entire part $[\cdot]$ can at most change the amount of ψ^* by one unit. Now, by (16.2) in theorem 16.1

$$|\psi^*(\tilde{n}';\omega_o^*) - \psi^*(n_o;\omega_o^*)| \leq 60 \cdot 2^r \quad \text{and} \quad |\psi^*(\tilde{n}_1;\tilde{\omega}_1^*) - \psi^*(\tilde{n}';\tilde{\omega}_1^*)| \leq 60 \cdot 2^\ell \ ,$$

where $\ell < r$, so a rough estimate gives

$$|\psi^*(\tilde{n}_1;\omega_o^*) - \psi^*(n_o;\omega_o^*)| \leq 60 \cdot \sum_{j=1}^{r} 2^j < 120 \cdot 2^r \ ,$$

proving iii) , and we have shown that $\Sigma \neq \emptyset$.

If we choose $(\bar{n}, \bar{\omega}^*, k) \in \Sigma$ such that k is the smallest possible integer, we only have to prove the inequality (16.13) and the claims concerning $\Omega(\bar{p}^*;k)$.

Suppose that (16.13) is not fulfilled, i.e. assume that $C^*(\bar{p}^*) \geq 2^{-k-1)} y$. Using lemma 11.3 once more we can find an integer $\ell \in \{1,2, \ldots, k-1\}$ such that

$$2^{-\ell} y \leq C^*(\bar{p}^*) < 2^{-(\ell-1)} y .$$

If $\bar{p}^* \subset G^*_{\ell L}$ we conclude that $(\bar{n},\bar{\omega}^*,\ell) \in \Sigma$, contradicting the definition of k, so we must have $\bar{p}^* \notin G^*_{\ell L}$. But in that case we use theorem 16.1 once more with ω^*_o, n_o, ℓ replaced by $\bar{\omega}^*, \bar{n}, \ell$. This gives us a new couple $(\tilde{n}_1, \tilde{\omega}^*_1)$, such that [cf. the construction above] $(\tilde{n}_1, \tilde{\omega}^*_1, \ell) \in \Sigma$, and since $\ell \leq k-1$, this is again contradicting the definition of k. Hence we have proved (16.13).

From the construction above follows that the splitting $\Omega(p^*;k)$ of $\bar{\omega}^*$ introduced in § 14 is defined. Let $\bar{\omega}^*(x)$ be the central interval with respect to x and $\Omega(\bar{p}^*;k)$ and let

$$\bar{p}^*(x) = (\psi^*(\bar{n};\bar{\omega}^*(x)), \omega^*(x)) .$$

Then of course x belongs to the middle half of both ω^*_o and $\bar{\omega}^*(x)$, and if x is not an endpoint of one of the two dyadic intervals in the middle half of ω^*_o, then $\bar{\omega}^*(x) \neq \omega^*_o$. By the construction of $\bar{\omega}^*(x)$ in § 14 this means that either $\bar{\omega}^*(x) \supset \omega^*_o$ or $\bar{\omega}^*(x) \subset \omega^*_o$ (strictly).

Suppose that $\bar{\omega}^*(x) \supset \omega^*_o$. Then $m(\bar{\omega}^*(x)) > 2 \cdot 2\pi \cdot 2^{-N}$, which by (14.11) in lemma 14.3 implies that

$$C^*(\bar{p}^*(x)) = C^*_{\psi^*(\bar{n};\bar{\omega}^*(x))}(\bar{\omega}^*(x);\chi^o_F) \geq 2^{-(k-1)} y .$$

Hence (lemma 11.3) there exists an integer $\ell \in \{1,2, \ldots, k-1\}$, such that

$$2^{-\ell} y \leq C^*(\bar{p}^*(x)) < 2^{-(\ell-1)} y .$$

If $\bar{p}^*(x) \in G^*_{\ell L}$, then $(\bar{n},\bar{\omega}^*(x),\ell) \in \Sigma$, contradicting the fact that k is minimal, so $\bar{p}^*(x) \notin G^*_{\ell L}$. But in that case we use theorem 16.1 once more with ω^*_o, n_o, r replaced by $\bar{\omega}^*(x), \bar{n}, \ell$, constructing another element $(\hat{n},\hat{\omega}^*,\ell) \in \Sigma$, which again is contradicting the minimality of k. Hence we conclude that $\bar{\omega}^*(x) \subset \omega^*_o$ (strictly).

Finally, since $\bar{\omega}^*(x)$ is a union of intervals from $\Omega(\bar{p}^*,k)$ we only have to concern ourselves that ω_o^* is a union of intervals from $\Omega(\bar{p}^*;k)$. This follows from the construction of $\bar{\omega}^*(x)$, from (16.13) and from the fact that $\bar{\omega}^*(x) \subset \omega_o^*$, i.e. $\bar{\omega}^*(x)\setminus\omega_o^* = \emptyset$. □

§ 17. Final estimate of $S_n^*(x;\chi_F^o;\omega_{-1}^*)$.

In this section we shall prove the estimate of $S_n^*(x;\chi_F^o;\omega_{-1}^*)$ for $x \in]-\pi,\pi[\setminus E_N$ and $|n| \leqq N$, which was claimed in the introduction to this chapter. The proof, which depends heavily on the preceeding sections, is very technical, and it uses an iterative procedure explained in lemma 17.3 below. Once this lemma has been proved the rest is easy.

Our preliminary goal is to define a *finite* sequence $\omega_{-1}^*, \omega_o^*, \omega_1^*, \dots, \omega_J^*$ of smoothing intervals and corresponding finite sequences of non-negative integers

$$n = n_{-1}, n_o, n_1, \dots, n_J = 0, \qquad 0 \leqq n_j \leqq N,$$

$$m_{-1} > m_o > m_1 > \dots > m_J,$$

such that

$$\left|S_{n_j}^*(x;\chi_F^o;\omega_j^*)\right| = \left|S_{n_{j+1}}^*(x;\chi_F^o;\omega_{j+1}^*)\right| + O\left(L\cdot m_j\cdot 2^{1-m_j}\cdot y\right), \quad j = -1,0,1,\dots,J-1.$$

Since the n-sequence is not necessarily decreasing, we shall for technical reasons prove the existence of another finite sequence $k_{-1}, k_o, k_1, \dots, k_J$ consisting of non-negative integers, such that

$$k_{j+1} < m_j \leqq k_j \qquad \text{and} \qquad n_{j+1} \leqq \left(1+2^{-k_j}\right)n_j, \qquad j = -1,0,1,\dots,J-1.$$

Let us first examine one particular ω_j^* and let $n_j \in N_o$ be the corresponding integer. If ω_j^* is composed of two dyadic intervals from level ν, and $n \in N_o$ satisfies the condition

$$\psi^*(n;\omega_j^*) = n_j \cdot 2^{-\nu-1},$$

then of course

(17.1) $\quad 0 \leqq n - n_j < 2^{\nu+1} \quad$ and $\quad \psi^*(n;\omega_j^*) = \psi^*(n_j;\omega_j^*)$.

If on the other hand $n \in \mathbb{N}$ is given, we simply define n_j by

$$n_j = \frac{4 \cdot 2\pi}{m(\omega_j^*)} \cdot \psi^*(n;\omega_j^*) , \tag{17.2}$$

and (n, n_j, ω_j^*) has the properties described above in (17.1).

If $n_j \in \mathbb{N}_o$ is given and we want (17.1) to be fulfilled for $n = n_j$, then of course $\psi^*(n_j;\omega_j^*)$ defines a level number $\nu \in \mathbb{N}_o$, such that ω_j^* is composed of two dyadic intervals from level ν, and $\psi^*(n_j;\omega_j^*) = n_j \cdot 2^{-\nu-1}$. Especially, $n_j = 0$ if and only if $\psi^*(n_j;\omega_j^*) = 0$. Using that x should belong to the middle half of ω_j^* we have only got two choices of ω_j^* for a given n_j. Let ω_j^* be any one of these.

Lemma 17.1. *Let $N \geqq 7$. If $n_j \neq 0$, then $m(\omega_j^*) > 8 \cdot 2\pi \cdot 2^{-N}$.*

Proof. Since $\sum_{i=1}^{+\infty} \log(1+2^{-i}) < \sum_{i=1}^{+\infty} 2^{-i} = 1$, we have $\prod_{i=1}^{+\infty} (1+2^{-i}) < e$. If $n_{-1} = n$, then the conditions $k_{j+1} < k_j$ and $n_{j+1} \leqq (1+2^{-k_j})n_j$ imply for $N \geqq 7$ that

$$n_j \leqq \prod_{i=1}^{+\infty} (1+2^{-i})N < e\,N \leqq \frac{1}{6} \cdot 2^N ,$$

so if $m(\omega_j^*) \leqq 8 \cdot 2\pi \cdot 2^{-N}$, then

$$\psi^*(n_j;\omega_j^*) \cdot 2^{N+1} \leqq \left[\frac{n_j}{8\pi} \cdot m(\omega_j^*)\right] \cdot 2^{N+1} \leqq \left[\frac{n_j}{8\pi} \cdot 8 \cdot 2\pi \cdot 2^{-N}\right] \cdot 2^{N+1} \leqq 4n_j \leqq \frac{2}{3} \cdot 2^N ,$$

so $\psi^*(n_j;\omega_j^*) \leqq \frac{1}{3}$, and as $\psi^*(n_j;\omega_j^*) \in \mathbb{N}_o$ we get $\psi^*(n_j;\omega_j^*) = 0$. According to the remarks above this means that $n_j = 0$, and the lemma follows. □

If $n_{-1} = n$ and $\omega_{-1}^* =]-4\pi, 4\pi]$ are given, we get from lemma 11.3 that there exists an integer $k \in \mathbb{N}$, such that

$$2^{-k} y \leqq C_n^*(\omega_{-1}^*;\chi_F^o) < 2^{-(k-1)} y . \tag{17.3}$$

Lemma 17.2. *If* $k \in \mathbb{N}$ *is defined by (17.3) and* $]-2\pi, 2\pi] \not\subseteq S^*$ *, then*

$$(\psi^*(n_{-1};\omega^*_{-1}), \omega^*_{-1}) \in G^*_{kL} \ .$$

Proof. In § 9 we defined $\psi^*(n;\omega^*_{-1}) = n$, so we might as well write $(n_{-1},\omega^*_{-1}) \in G^*_{kL}$. Let us suppose that $(\psi^*(n_{-1};\omega^*_{-1}), \omega^*_{-1}) \notin G^*_{kL}$. Let $m \in \mathbb{N}$ be any number, such that $|m-n_{-1}| = |m-n| = |m-\psi^*(n;\omega^*_{-1})| < 2^{10\,kL}$. Then $(m;\omega_{r0}) \notin G_{kL}$, and we get from the definition (12.1) of $G_k(\omega_{r0})$ that

$$(17.4) \qquad |c_m(\omega_{r0};\overset{o}{\chi}_F)| < 2^{-kL}\, y^{p/2} \ .$$

Since $]-2\pi, 2\pi] \not\subseteq S^*$, we have $\omega_{r0} \not\subseteq S$, and it follows from (11.2) that

$$(17.5) \qquad \left\{\frac{1}{m(\omega_{r0})}\int_{\omega_{r0}} |\overset{o}{\chi}_F(x)|^2 dx\right\}^{\frac{1}{2}} = \left\{\frac{1}{m(\omega_{r0})}\int_{\omega_{r0}} \overset{o}{\chi}_F(x)\, dx\right\}^{\frac{1}{2}} < y^{p/2} \ .$$

Now, (17.5) and (17.4) are exactly the assumptions of lemma 9.6, so if we choose $A = y^{p/2}$ and $B = 2^{-kL}\, y^{p/2}$ and $M = 2^{10\,kL}$, we get from that lemma

$$c_n(\omega_{r0};\overset{o}{\chi}_F) \leqq 9\left\{2^{-kL}\, y^{p/2} \cdot 10\,kL \cdot \log 2 + y^{p/2} \cdot 2^{-5\,kL}\right\} < 2^{-kL/2}\, y < 2^{-k}\, y \ ,$$

where we have used that $L = L(p) \geqq 100$, cf. (11.6). But this inequality is contradicting the definition (17.3) of k . □

We shall now turn to the crucial lemma, in which we shall make precise the roles of the finite sequences mentioned above.

Lemma 17.3. *Let* $N \geqq 7$, $1 < p < +\infty$ *and* $y \in \mathbb{R}_+$ *be given constants, and let* $F \subseteqq\ ]-\pi, \pi]$ *be a given measurable set. Then there exists another measurable set* $E \subseteqq\ ]-4\pi, 4\pi]$ *, such that*

$$m(E) \leqq C_p^p\, y^{-p}\, m(F) \ ,$$

where the set E *furthermore satisfies the following condition:*

To each $x \in\ ]-\pi, \pi]\backslash E$ *and each* $n \in \{0, 1, \ldots, N\}$ *one can find a* finite *sequence*

$$\omega^*_{-1}\, ,\ \omega^*_{0}\, ,\ \omega^*_{1}\, ,\ \ldots\, ,\ \omega^*_{J}$$

of smoothing intervals, and three corresponding sequences

$n = n_{-1}, n_o, n_1, \ldots, n_J = 0$, $m_{-1}, m_o, m_1, \ldots, m_J$, $k_{-1}, k_o, k_1, \ldots, k_J$,

of non-negative integers, such that

$$n_j = \frac{4\cdot 2\pi}{m(\omega_j^*)} \cdot \psi^*(n_j;\omega_j^*) , \quad n_{j+1} \leqq (1+2^{-k_j})n_j , \quad k_{j+1} < m_j \leqq k_j ,$$

and such that

$$(17.6) \qquad |S^*_{n_j}(x;\chi^o_F;\omega^*_j)| \leqq |S^*_{n_{j+1}}(x;\chi^o_F;\omega^*_{j+1})| + c_{12}\cdot L\cdot m_j \cdot 2^{-(m_j-1)} y$$

and

$$(17.7) \qquad |S^*_o(x;\chi^o_F;\omega^*_j)| \leqq L\cdot y .$$

Here, $L = L(p) \geqq 100$ *is the constant introduced in (11.6), and* c_{12} *is a constant, which is independent of* N, p, y *and* F *. (It is possible to prove that* c_{12} *may be chosen equal to* $2\cdot 10^4$ *.)*

Proof. We may assume that $m(F) > 0$. Let $n_{-1} = n$ and $\omega^*_{-1} =]-4\pi, 4\pi]$, and let $k \in \mathbb{N}$ be defined by (17.3), i.e.

$$2^{-k} y \leqq C^*_{n_{-1}}(\omega^*_{-1};\chi^o_F) < 2^{-(k-1)} y .$$

By lemma 17.2 we have $(\psi^*(n_{-1};\omega^*_{-1}), \omega^*_{-1}) \in G^*_{kL}$, so the splitting $\Omega((\psi^*(n;\omega^*_{-1}), \omega^*_{-1}); k)$ of ω^*_{-1} is well-defined.

We choose E as the exceptional set E_N defined by (15.12). Since $x \notin E$ it follows from theorem 15.7 that

$$(17.8) \qquad |S^*_{n_{-1}}(x;\chi^o_F;\omega^*_{-1})| \leqq |S^*_{n_{-1}}(x;\chi^o_F;\omega^*(x))| + c_9\cdot Lk\cdot 2^{-(k-1)} y .$$

We define

$$k_{-1} = m_{-1} = k , \qquad \omega^*_o = \omega^*(x) \qquad \text{and} \qquad n_o = \frac{4\cdot 2\pi}{m(\omega^*_o)} \cdot \psi(n_{-1};\omega^*_o) .$$

If $\omega^*(x)$ is composed of two dyadic intervals from level ν, $\omega^*_o = \omega^*(x) = \omega_{j\nu} \cup \omega_{j+1,\nu}$, it follows from (17.2) and (17.1) that

$0 \leqq n_{-1} - n_o < 2^{\nu+1}$, so we may use lemma 10.3 to get

$$\left|S^*_{n_{-1}}(x;\chi^o_F;\omega^*_o)\right|$$

$$\leqq \left|S^*_{n_o}(x;\chi^o_F;\omega^*_o)\right| + c_4 \cdot \max\left\{C_{\psi(n_o;\omega_{j\nu})}(\omega_{j\nu};\chi^o_F)\ ,\ C_{\psi(n_o;\omega_{j+1,\nu})}(\omega_{j+1,\nu};\chi^o_F)\right\} .$$

Now, by definition of $\omega^*(x)$ in § 14, at least one of the intervals $\omega_{j\nu}$ and $\omega_{j+1,\nu}$ belongs to the splitting, so by (14.10) in lemma 14.3 we get

$$\left|S^*_{n_{-1}}(x;\chi^o_F;\omega^*_o)\right| \leqq \left|S^*_{n_o}(x;\chi^o_F;\omega^*_o)\right| + c_4 \cdot 2^{-(k-1)}\, y$$

$$\leqq \left|S^*_{n_o}(x;\chi^o_F;\omega^*_o)\right| + 10^{-2}\, c_4 \cdot Lk \cdot 2^{-(k-1)}\, y \ ,$$

as $Lk \geqq 100$. When this inequality is substituted into (17.8) we find

$$\left|S^*_{n_{-1}}(x;\chi^o_F;\omega^*_{-1})\right| \leqq \left|S^*_{n_o}(x;\chi^o_F;\omega^*_o)\right| + c_{13} \cdot Lk \cdot 2^{-(k-1)}\, y \ . \tag{17.9}$$

If $n_o = 0$ the process stops. If $n_o \neq 0$ it follows from lemma 17.1 that $m(\omega^*_o) > 8 \cdot 2\pi \cdot 2^{-N}$, so we infer from (14.4) in lemma 14.1 that either

$$C^*_{\psi^*(n_o;\omega^*_o)}(\omega^*_o;\chi^o_F) \geqq 2^{-(k-1)}\, y \ , \tag{17.10}$$

or $\chi^o_F \cdot \chi_{\omega^*_o} \equiv 0$. In the latter case $S^*_{n_o}(x;\chi^o_F;\omega^*_o) = 0$ and (17.6) and (17.7) become trivial for ω^*_o, so we may assume (17.10). Then by lemma 11.3 we conclude that there exists a $k_o \in N$, such that

$$2^{-k_o}\, y \leqq C^*_{\psi^*(n_o;\omega^*_o)}(\omega^*_o;\chi^o_F) < 2^{-(k_o-1)}\, y \ ,$$

and it follows at once from (17.10) that $k_o < k = m_{-1}$.

We define $p^*_o = (\psi^*(n_o;\omega^*_o),\ \omega^*_o)$.

If $p^*_o \in G^*_{k_oL}$, then the splitting $\Omega(p^*_o;k_o)$ of ω^*_o is defined, and using the same procedure as above in the derivation of (17.10) [applying lemma 10.3 and lemma 14.3] we get

$$(17.11)\qquad \left|S^*_{n_o}(x;\chi^o_F;\omega^*_o)\right| \leq \left|S^*_{n_1}(x;\chi^o_F;\omega^*_1)\right| + c_{13}\cdot Lm_o\cdot 2^{-(m_o-1)}y ,$$

where we have put

$$\omega^*_1 = \omega^*_o(x) , \qquad m_o = k_o \qquad \text{and} \qquad n_1 = \frac{4\cdot 2\pi}{m(\omega^*_1)}\cdot \psi^*(n_o;\omega^*_1) .$$

If $p_o \notin G^*_{k_oL}$ and $\psi^*(n_o;\omega_o) > 120\cdot 2^{2k_o}$, then choose $\tilde{n}$ according to theorem 16.1 and $\bar{n}, \bar{\omega}^*, r$ according to theorem 16.2. Especially, $r \leq k_o$. Using theorem 15.7 we get

$$(17.12)\qquad \left|S^*_{\bar{n}}(x;\chi^o_F;\omega^*_o)\right| \leq \left|S^*_{\bar{n}}(x;\chi^o_F;\bar{\omega}^*(x))\right| + c_9\cdot Lr\cdot 2^{-(r-1)}y$$

As $|\psi^*(\bar{n};\omega^*_o) - \psi^*(n_o;\omega^*_o)| < 120\cdot 2^{k_o} < 120\cdot 2^{2k_o}$, it follows from theorem 16.1 that

$$(17.13)\qquad \left|S^*_{n_o}(x;\chi^o_F;\omega^*_o)\right| \leq \left|S^*_{\bar{n}}(x;\chi^o_F;\omega^*_o)\right| + 200\cdot\left\{C^*_{\psi^*(\tilde{n};\omega^*_o)}(\omega^*_o;\chi^o_F) + 2^{-(k_o-1)}y\right\} .$$

On the other hand we get from theorem 16.2 that

$$C^*_{\psi^*(\tilde{n};\omega^*_o)}(\omega^*_o;\chi^o_F) < 2^{-(r-1)}y ,$$

so substituting this inequality and (17.12) into (17.13) and using that $Lr \geq 100$, we get

$$(17.14)\qquad \left|S^*_{n_o}(x;\chi^o_F;\omega^*_o)\right| \leq \left|S^*_{\bar{n}}(x;\chi^o_F;\bar{\omega}^*_o(x))\right| + c_{14}\cdot Lr\cdot 2^{-(r-1)}y .$$

In this case we put

$$(17.15)\qquad \omega^*_1 = \bar{\omega}^*(x) , \qquad m_o = m , \qquad \text{and} \qquad n_1 = \frac{4\cdot 2\pi}{m(\omega^*_1)}\cdot \psi^*(\bar{n};\omega^*_1) .$$

We have not concluded the construction in this case, however, since n_1 need not be equal to $\bar{n}$, and $\bar{n}$ may be larger than n_o . We shall prove

<u>Lemma 17.4.</u> *Let* $n_1 = \frac{4\cdot 2\pi}{m(\omega_1^*)}\cdot\psi^*(\bar n;\omega_1^*)$ *and* $\psi^*(n_o;\omega_o^*) > 120\cdot 2^{2k_o}$ *. Then*

$$n_1 \leqq \bar n \leqq (1+2^{-k_o})n_o \ .$$

<u>Proof.</u> If $\bar n \leqq n_o$, the claim is trivial, so assume that $n_o < \bar n$. If ω_o^* is composed of two intervals from level ν , we get by assumption

$$\bar n\cdot 2^{-(\nu+1)} - n_o\cdot 2^{-(\nu+1)} < 120\cdot 2^{k_o} = 2^{-k_o}\cdot 120\cdot 2^{2k_o} < 2^{-k_o}\cdot n_o\cdot 2^{-(\nu+1)} ,$$

from which follows by a rearrangement that $\bar n < (1+2^{-k_o})n_o$. □

Using this lemma and (14.10) in lemma 14.3 and corollary 10.6 we get from (17.14), using once more that $Lr \geqq 100$,

$$(17.16)\qquad \left|S^*_{n_o}(x;\chi_F^o;\omega_o^*)\right| \leqq \left|S^*_{\bar n}(x;\chi_F^o;\omega_1^*)\right| + c_{14}\cdot Lr\cdot 2^{-(r-1)}y$$

$$\leqq \left|S^*_{n_1}(x;\chi_F^o;\omega_1^*)\right| + c_4\cdot\max\left\{C_{m_o}(\omega_{j\nu};\chi_F^o),C_{m_o}(\omega_{j+1,\nu};\chi_F^o)\right\} + c_{14}\cdot Lr\cdot 2^{-(r-1)}y$$

$$\leqq \left|S^*_{n_1}(x;\chi_F^o;\omega_1^*)\right| + c_{15}\cdot Lr\cdot 2^{-(r-1)}y ,$$

i.e. we have proved (17.6) in this case.

At last, consider the case where $p_o^* \notin G^*_{k_oL}$ and $\psi^*(n_o;\omega_o^*) \leqq 120\cdot 2^{k_o}$. Again, choose $\tilde n$ as in theorem 16.1 and $\bar n , \bar\omega^* , r$ as in theorem 16.2. Then by theorem 16.1 we get for $n=0$,

$$(17.17)\qquad \left|S^*_{n_o}(x;\chi_F^o;\omega_o^*)\right| \leqq \left|S^*_o(x;\chi_F^o;\omega_o^*)\right| + 200\cdot\left\{C^*_{\psi^*(\tilde n;\omega_o^*)}(\omega_o^*;\chi_F^o) + 2^{-(k_o-1)}y\right\}$$

so using (16.13) in theorem 16.2 we get

$$\left|S^*_{n_o}(x;\chi_F^o;\omega_o^*)\right| \leqq \left|S^*_o(x;\chi_F^o;\omega_o^*)\right| + 200\left\{2^{-(r-1)}y + 2^{-(k_o-1)}y\right\} .$$

Now, $\omega_o^* = \omega^*(x)$ is an interval of σ_x-type, and as $x \notin V_L^*$ [cf. (11.8)], i.e. $\sup_{\sigma_x}\left|\frac{1}{\pi}\,(pv)\int_{\sigma_x}\frac{\overset{o}{\chi}_F(t)}{x-t}dt\right| \leqq L\,y$, we get

$$\left|S_o^*(x;\overset{o}{\chi}_F;\omega_o^*)\right| = \left|\frac{1}{\pi}(pv)\int_{\omega_o^*}\frac{\overset{o}{\chi}_F(t)}{x-t}dt\right| \leqq \sup_{\sigma_x}\left|\frac{1}{\pi}(pv)\int_{\omega_o^*}\frac{\overset{o}{\chi}_F(t)}{x-t}dt\right| \leqq L\,y\,,$$

proving (17.7), [as ω_o^* may be replaced by ω_j^* , whenever ω_j^* is defined]. In this case we put $m_o = m = 1$ and $n_1 = 0$ and $\omega_o^* = \omega_1^*$, and we get

$$(17.18)\qquad \left|S_{n_o}^*(x;\overset{o}{\chi}_F;\omega_o^*)\right| \leqq \left|S_o^*(x;\overset{o}{\chi}_F;\omega_o^*)\right| + c_{12}\cdot Lm_o\cdot 2^{-(m_o-1)}\,y\,.$$

The whole process is then iterated until either (17.18) is obtained, or until we get so small intervals in (17.14) or (17.16) that we by lemma 17.1 conclude that $n_J = 0$. One of these possibilities will occur within a finite number of steps, and lemma 17.3 is proved. □

Finally we have the following theorem

Theorem 17.5. *Let* $N \in \mathbb{N}$, $N \geqq 7$, $p \in\,]1,+\infty[$, $y \in \mathbb{R}_+$ *and the measurable set* $F \subseteqq\,]-\pi,\pi[$ *be given. There exists a measurable set* $E \subseteqq\,]-4\pi, 4\pi]$, *such that*

$$m(E) \leqq C_p^p\, y^{-p}\, m(F)\,,$$

and such that we have for every $n \in \mathbb{Z}$ *with* $|n| \leqq N$ *and for every* $x \in\,]-\pi,\pi]\backslash E$

$$\left|S_n^*(x;\overset{o}{\chi}_F;\omega_{-1}^*)\right| \leqq 5\,c_{12}\cdot L\,y\,,$$

where C_p *and* $L = L(p)$ *and* c_{12} *are the constants given in lemma 17.3.*

Proof. Case 1. $(n = 0)$. By lemma 17.3 we have

$$|S_o^*(x;\overset{o}{\chi}_F;\omega_{-1}^*)| \leqq c_{12}\cdot L\,y\,.$$

Case 2. $(0 < n \leq N)$. By lemma 17.3 we have

$$\left|S_n^*(x;\chi_F^o;\omega_{-1}^*)\right| = \left|S_{n_{-1}}^*(x;\chi_F^o;\omega_{-1}^*)\right|$$

$$\leq \left|S_{n_o}^*(x;\chi_F^o;\omega_o^*)\right| + c_{12} \cdot L\,y \cdot m_{-1} \cdot 2^{-(m_{-1}-1)} \leq \dots$$

$$\leq \left|S_{n_J}^*(x;\chi_F^o;\omega_J^*)\right| + c_{12} \cdot L\,y \cdot \sum_{j=-1}^{J-1} m_j \cdot 2^{-(m_j-1)}$$

$$\leq c_{12} \cdot L\,y + c_{12} \cdot L\,y \cdot \sum_{i=1}^{+\infty} i \cdot 2^{-(i-1)} = 5\,c_{12}\;L\,y\,,$$

where we have used that $k_{j+1} < m_j \leq k_j$, so especially all the m_j are different.

Case 3. $(-N \leq n < 0)$. As $|S_n^*(x;\chi_F^o;\omega^*)| = |S_{-n}^*(x;\chi_F^o;\omega^*)|$ this case follows immediately from case 2. □

§ 18. Proof of theorem 4.2.

In § 4 we proved the Carleson-Hunt theorem assuming that theorem 4.2 is true. All the results in the sections following § 4 have been proved in order to be able here to prove theorem 4.2, thus concluding the proof of the famous Carleson-Hunt theorem.

In § 10 we introduced the operator $M^* : L^p([-\pi,\pi]) \to \mathfrak{M}$ by

(18.1) $$M^*f(x) = \sup_{n\in Z} |S_n^*(x;f;\omega_{-1}^*)| = \sup_{n\in N_o} |S_n^*(x;f;\omega_{-1}^*)|\,.$$

Concerning this operator we prove

Theorem 18.1. *The operator* M^* *is of restricted weak type* p, $p \in \,]1,+\infty[$.

Proof. We recall from § 1 that we have to prove the existence of a constant A_p, such that

$$(18.2) \qquad m(\{x \in]-\pi,\pi] \mid M^*\chi_F(x) > y\}) \leqq A_p^p\, y^{-p}\, m(F)$$

for every $y \in \mathbb{R}_+$, every $p \in]1,+\infty[$ and every measurable set $F \subseteq]-\pi,\pi]$.

Let y, p and F be given as above and put

$$E_N^* = \left\{x \in]-\pi,\pi] \mid \sup_{|n| \leqq N} |S_n^*(x;\chi_F^o;\omega_{-1}^*)| > y\right\}, \qquad N \in \mathbb{N},$$

$$E^* = \left\{x \in]-\pi,\pi] \mid M^*\chi_F(x) > y\right\}.$$

Clearly, $E_N^* \uparrow E^*$, and in order to prove (18.2) it therefore suffices to prove the existence of a constant A_p, such that

$$m(E_N^*) \leqq A_p^p\, y^{-p}\, m(F) \qquad \text{for all } N \in \mathbb{N}.$$

By the definition of E_N^* we have

$$\sup_{|n| \leqq N} |S_n^*(x;\chi_F^o;\omega_{-1}^*)| \leqq y \qquad \text{for } x \in]-\pi,\pi] \setminus E_N^*.$$

We now put $y = 5\,c_{12}\,L\,y_1$ (i.e. $y_1 = \frac{1}{5} \cdot c_{12}^{-1} \cdot L^{-1} y$), where c_{12} and L are the constants from theorem 17.5. If E_N denotes the set in theorem 17.5 corresponding to y_1 and N above, we have

$$]-\pi,\pi] \setminus E_N^* \supseteq]-\pi,\pi] \setminus E_N,$$

i.e. $E_N^* \subseteq E_N$, and we infer from theorem 17.5 that

$$m(E_N^*) \leqq m(E_N) \leqq C_p^p\, y_1^{-p}\, m(F) = (5\,c_{12} \cdot C_p)^p \cdot y^{-p}\, m(F),$$

which proves the theorem with $A_p = 5\,c_{12} \cdot C_p$. □

Corollary 18.2. *The operator* M^* *is of restricted type* p, $p \in]1,+\infty[$.

Proof. This follows immediately from theorem 18.1 and lemma 1.8 (interpolating between p_o and p_1 where $1 < p_o < p < p_1 < +\infty$). □

Corollary 18.3. *The operator* M *is of restricted type* p, $p \in]1,+\infty[$. *More precisely, to each* $p \in]1,+\infty[$ *there exists a constant* A_p , *such that* $\|M\chi_E\|_p \leqq A_p\|\chi_E\|_p$ *for all measurable sets* $E \subseteqq]-\pi,\pi]$.

Proof. This follows immediately from theorem 10.2 and corollary 18.2. □

Let $N \in \mathbb{N}$ be fixed and put

$$M_N f(x) = \max_{0 \leqq n \leqq N} |S_n(x;f)| .$$

Let $\alpha :]-\pi,\pi] \to \{0,1,2,\ldots,N\}$ be a simple function. When the range of α is contained in $\{0,1,2,\ldots,N\}$, we call α a *simple function of order* N .

We next define an operator T_α on L^p by

$$(18.3) \qquad T_\alpha f(x) = S_{\alpha(x)}(x;f) , \quad x \in]-\pi,\pi] .$$

Clearly, T_α is a linear operator on L^p for every simple function α of order N .

Lemma 18.4. *Let* α *denote any simple function of order* N . *Then we have for any measurable set* $E \subseteqq]-\pi,\pi]$ *and any* $p \in]1,+\infty[$,

$$\|T_\alpha \chi_E\|_p \leqq A_p\|\chi_E\|_p ,$$

where A_p *is the constant in corollary 18.3.*

Proof. As $Mf(x) = \sup_{n \in \mathbb{N}_0} |S_n(x;f)|$ we clearly have $|T_\alpha f(x)| \leqq Mf(x)$ and hence also $\|T_\alpha f\|_p \leqq \|Mf\|_p$. The result now follows from corollary 18.3.

□

Lemma 18.5. *To every* $p \in]1,+\infty[$ *there exists a constant* C_p' , *such that for every simple function* $f \in L^p(]-\pi,\pi])$ *and for every simple function* α *of order* N *we have* $\|T_\alpha f\|_p \leqq C_p'\|f\|_p$.

Proof. This follows immediately from lemma 18.4 and the theorem of Stein-Weiss in § 3 (interpolating between p_0 and p_1, where $1 < p_0 < p < p_1 < +\infty$). We note that the constant C'_p only depends on p and not on α, cf. the Marcinkiewicz interpolation theorem. □

Lemma 18.6. *For every* $f \in L^p(]-\pi, \pi])$ *and every simple function* α *of order* N *we have*

$$\|T_\alpha f\|_p \leq C'_p \|f\|_p ,$$

i.e. T_α *is of type* p, $p \in]1,+\infty[$. *Here* C'_p *is the constant in lemma 18.5.*

Proof. Let $f \in L^p(]-\pi, \pi])$. It is well-known that there exists a sequence of simple functions f_k satisfying

$$f_k(x) \to f(x) , \qquad |f_k(x)| \leq |f(x)| \qquad \text{for all } x \in]-\pi, \pi] .$$

Clearly, we also have $|T_\alpha f_k(x)| \to |T_\alpha f(x)|$ for any simple function α of order N and $\|f_k\|_p \to \|f\|_p$ by Lebesgue's theorem on dominated convergence.

Using Fatou's lemma and lemma 18.5 we have

$$\|T_\alpha f\|_p \leq \liminf_{k\to+\infty} \|T_\alpha f_k\|_p \leq C'_p \lim_{k\to+\infty} \|f_k\|_p = C'_p \|f\|_p . \qquad \square$$

Finally, we are able to prove

Theorem 4.2. *The operator* M *is of type* p *for all* $p \in]1,+\infty[$.

Proof. Let $f \in L^p(]-\pi, \pi])$. From the definition of the operator M_N it follows that there exists a simple function α_0 of order N (where α_0 depends on f), such that

$$|T_{\alpha_0} f(x)| = \max_{0\leq n\leq N} |S_n(x;f)| = M_N f(x) \qquad \text{for all } x \in]-\pi, \pi] .$$

Then, using lemma 18.6, we have $\|M_N f\|_p \leq C'_p \|f\|_p$. As the sequence $M_N f(x)$ is increasing and $\lim_{N\to+\infty} M_N f(x) = M f(x)$ for every $x \in]-\pi, \pi]$, the theorem follows. □

REFERENCES.

[1] Carleson, L. Convergence and growth of partial sums of Fourier series. Acta Math. 116 (1966), 135-157.

[2] Garsia, A. Topics in almost everywhere convergence. Markham Publishing Company, Chicago, 1970.

[3] de Guzmán, M. Differentiation of Integrals in R^n. Lecture Notes in Mathematics 481, Springer, 1975.

[4] Hunt, R.A. On the convergence of Fourier series. Orthogonal Expansions and Their Continuous Analogues. Proc. Conf. Edwardsville. Ill. (1967), 235-255. Southern Illinois Univ. Press, Carbondale, Ill. (1968).

[5] Loomis. A note on the Hilbert transform. Bull. Am. Math. Soc. 52 (1946), 1082-1086.

[6] Marsden, J. Basic complex analysis. Freeman and Company, San Francisco, 1973.

[7] Mozzochi, Ch.J. On the Pointwise Convergence of Fourier Series. Lecture Notes in Mathematics 199, Springer 1971.

[8] Titchmarsh. Introduction to the theory of Fourier integrals. Clarendon Press, Oxford, 1962.

[9] Zygmund, A. Trigonometric series, Vol. I-II, Cambridge University Press, 1959.

INDEX.

Vol. 759: R. L. Epstein, Degrees of Unsolvability: Structure and Theory. XIV, 216 pages. 1979.

Vol. 760: H.-O. Georgii, Canonical Gibbs Measures. VIII, 190 pages. 1979.

Vol. 761: K. Johannson, Homotopy Equivalences of 3-Manifolds with Boundaries. 2, 303 pages. 1979.

Vol. 762: D. H. Sattinger, Group Theoretic Methods in Bifurcation Theory. V, 241 pages. 1979.

Vol. 763: Algebraic Topology, Aarhus 1978. Proceedings, 1978. Edited by J. L. Dupont and H. Madsen. VI, 695 pages. 1979.

Vol. 764: B. Srinivasan, Representations of Finite Chevalley Groups. XI, 177 pages. 1979.

Vol. 765: Padé Approximation and its Applications. Proceedings, 1979. Edited by L. Wuytack. VI, 392 pages. 1979.

Vol. 766: T. tom Dieck, Transformation Groups and Representation Theory. VIII, 309 pages. 1979.

Vol. 767: M. Namba, Families of Meromorphic Functions on Compact Riemann Surfaces. XII, 284 pages. 1979.

Vol. 768: R. S. Doran and J. Wichmann, Approximate Identities and Factorization in Banach Modules. X, 305 pages. 1979.

Vol. 769: J. Flum, M. Ziegler, Topological Model Theory. X, 151 pages. 1980.

Vol. 770: Séminaire Bourbaki vol. 1978/79 Exposés 525-542. IV, 341 pages. 1980.

Vol. 771: Approximation Methods for Navier-Stokes Problems. Proceedings, 1979. Edited by R. Rautmann. XVI, 581 pages. 1980.

Vol. 772: J. P. Levine, Algebraic Structure of Knot Modules. XI, 104 pages. 1980.

Vol. 773: Numerical Analysis. Proceedings, 1979. Edited by G. A. Watson. X, 184 pages. 1980.

Vol. 774: R. Azencott, Y. Guivarc'h, R. F. Gundy, Ecole d'Eté de Probabilités de Saint-Flour VIII-1978. Edited by P. L. Hennequin. XIII, 334 pages. 1980.

Vol. 775: Geometric Methods in Mathematical Physics. Proceedings, 1979. Edited by G. Kaiser and J. E. Marsden. VII, 257 pages. 1980.

Vol. 776: B. Gross, Arithmetic on Elliptic Curves with Complex Multiplication. V, 95 pages. 1980.

Vol. 777: Séminaire sur les Singularités des Surfaces. Proceedings, 1976-1977. Edited by M. Demazure, H. Pinkham and B. Teissier. IX, 339 pages. 1980.

Vol. 778: SK1 von Schiefkörpern. Proceedings, 1976. Edited by P. Draxl and M. Kneser. II, 124 pages. 1980.

Vol. 779: Euclidean Harmonic Analysis. Proceedings, 1979. Edited by J. J. Benedetto. III, 177 pages. 1980.

Vol. 780: L. Schwartz, Semi-Martingales sur des Variétés, et Martingales Conformes sur des Variétés Analytiques Complexes. XV, 132 pages. 1980.

Vol. 781: Harmonic Analysis Iraklion 1978. Proceedings 1978. Edited by N. Petridis, S. K. Pichorides and N. Varopoulos. V, 213 pages. 1980.

Vol. 782: Bifurcation and Nonlinear Eigenvalue Problems. Proceedings, 1978. Edited by C. Bardos, J. M. Lasry and M. Schatzman. VIII, 296 pages. 1980.

Vol. 783: A. Dinghas, Wertverteilung meromorpher Funktionen in ein- und mehrfach zusammenhängenden Gebieten. Edited by R. Nevanlinna and C. Andreian Cazacu. XIII, 145 pages. 1980.

Vol. 784: Séminaire de Probabilités XIV. Proceedings, 1978/79. Edited by J. Azéma and M. Yor. VIII, 546 pages. 1980.

Vol. 785: W. M. Schmidt, Diophantine Approximation. X, 299 pages. 1980.

Vol. 786: I. J. Maddox, Infinite Matrices of Operators. V, 122 pages. 1980.

Vol. 787: Potential Theory, Copenhagen 1979. Proceedings, 1979. Edited by C. Berg, G. Forst and B. Fuglede. VIII, 319 pages. 1980.

Vol. 788: Topology Symposium, Siegen 1979. Proceedings, 1979. Edited by U. Koschorke and W. D. Neumann. VIII, 495 pages. 1980.

Vol. 789: J. E. Humphreys, Arithmetic Groups. VII, 158 pages. 1980.

Vol. 790: W. Dicks, Groups, Trees and Projective Modules. IX, 127 pages. 1980.

Vol. 791: K. W. Bauer and S. Ruscheweyh, Differential Operators for Partial Differential Equations and Function Theoretic Applications. V, 258 pages. 1980.

Vol. 792: Geometry and Differential Geometry. Proceedings, 1979. Edited by R. Artzy and I. Vaisman. VI, 443 pages. 1980.

Vol. 793: J. Renault, A Groupoid Approach to C*-Algebras. III, 160 pages. 1980.

Vol. 794: Measure Theory, Oberwolfach 1979. Proceedings 1979. Edited by D. Kölzow. XV, 573 pages. 1980.

Vol. 795: Séminaire d'Algèbre Paul Dubreil et Marie-Paule Malliavin. Proceedings 1979. Edited by M. P. Malliavin. V, 433 pages. 1980.

Vol. 796: C. Constantinescu, Duality in Measure Theory. IV, 197 pages. 1980.

Vol. 797: S. Mäki, The Determination of Units in Real Cyclic Sextic Fields. III, 198 pages. 1980.

Vol. 798: Analytic Functions, Kozubnik 1979. Proceedings. Edited by J. Ławrynowicz. X, 476 pages. 1980.

Vol. 799: Functional Differential Equations and Bifurcation. Proceedings 1979. Edited by A. F. Izé. XXII, 409 pages. 1980.

Vol. 800: M.-F. Vignéras, Arithmétique des Algèbres de Quaternions. VII, 169 pages. 1980.

Vol. 801: K. Floret, Weakly Compact Sets. VII, 123 pages. 1980.

Vol. 802: J. Bair, R. Fourneau, Etude Géometrique des Espaces Vectoriels II. VII, 283 pages. 1980.

Vol. 803: F.-Y. Maeda, Dirichlet Integrals on Harmonic Spaces. X, 180 pages. 1980.

Vol. 804: M. Matsuda, First Order Algebraic Differential Equations. VII, 111 pages. 1980.

Vol. 805: O. Kowalski, Generalized Symmetric Spaces. XII, 187 pages. 1980.

Vol. 806: Burnside Groups. Proceedings, 1977. Edited by J. L. Mennicke. V, 274 pages. 1980.

Vol. 807: Fonctions de Plusieurs Variables Complexes IV. Proceedings, 1979. Edited by F. Norguet. IX, 198 pages. 1980.

Vol. 808: G. Maury et J. Raynaud, Ordres Maximaux au Sens de K. Asano. VIII, 192 pages. 1980.

Vol. 809: I. Gumowski and Ch. Mira, Recurences and Discrete Dynamic Systems. VI, 272 pages. 1980.

Vol. 810: Geometrical Approaches to Differential Equations. Proceedings 1979. Edited by R. Martini. VII, 339 pages. 1980.

Vol. 811: D. Normann, Recursion on the Countable Functionals. VIII, 191 pages. 1980.

Vol. 812: Y. Namikawa, Toroidal Compactification of Siegel Spaces. VIII, 162 pages. 1980.

Vol. 813: A. Campillo, Algebroid Curves in Positive Characteristic. V, 168 pages. 1980.

Vol. 814: Séminaire de Théorie du Potentiel, Paris, No. 5. Proceedings. Edited by F. Hirsch et G. Mokobodzki. IV, 239 pages. 1980.

Vol. 815: P. J. Slodowy, Simple Singularities and Simple Algebraic Groups. XI, 175 pages. 1980.

Vol. 816: L. Stoica, Local Operators and Markov Processes. VIII, 104 pages. 1980.

Vol. 817: L. Gerritzen, M. van der Put, Schottky Groups and Mumford Curves. VIII, 317 pages. 1980.

Vol. 818: S. Montgomery, Fixed Rings of Finite Automorphism Groups of Associative Rings. VII, 126 pages. 1980.

Vol. 819: Global Theory of Dynamical Systems. Proceedings, 1979. Edited by Z. Nitecki and C. Robinson. IX, 499 pages. 1980.

Vol. 820: W. Abikoff, The Real Analytic Theory of Teichmüller Space. VII, 144 pages. 1980.

Vol. 821: Statistique non Paramétrique Asymptotique. Proceedings, 1979. Edited by J.-P. Raoult. VII, 175 pages. 1980.

Vol. 822: Séminaire Pierre Lelong–Henri Skoda, (Analyse) Années 1978/79. Proceedings. Edited by P. Lelong et H. Skoda. VIII, 356 pages, 1980.

Vol. 823: J. Král, Integral Operators in Potential Theory. III, 171 pages. 1980.

Vol. 824: D. Frank Hsu, Cyclic Neofields and Combinatorial Designs. VI, 230 pages. 1980.

Vol. 825: Ring Theory, Antwerp 1980. Proceedings. Edited by F. van Oystaeyen. VII, 209 pages. 1980.

Vol. 826: Ph. G. Ciarlet et P. Rabier, Les Equations de von Kármán. VI, 181 pages. 1980.

Vol. 827: Ordinary and Partial Differential Equations. Proceedings, 1978. Edited by W. N. Everitt. XVI, 271 pages. 1980.

Vol. 828: Probability Theory on Vector Spaces II. Proceedings, 1979. Edited by A. Weron. XIII, 324 pages. 1980.

Vol. 829: Combinatorial Mathematics VII. Proceedings, 1979. Edited by R. W. Robinson et al.. X, 256 pages. 1980.

Vol. 830: J. A. Green, Polynomial Representations of GL_n. VI, 118 pages. 1980.

Vol. 831: Representation Theory I. Proceedings, 1979. Edited by V. Dlab and P. Gabriel. XIV, 373 pages. 1980.

Vol. 832: Representation Theory II. Proceedings, 1979. Edited by V. Dlab and P. Gabriel. XIV, 673 pages. 1980.

Vol. 833: Th. Jeulin, Semi-Martingales et Grossissement d'une Filtration. IX, 142 Seiten. 1980.

Vol. 834: Model Theory of Algebra and Arithmetic. Proceedings, 1979. Edited by L. Pacholski, J. Wierzejewski, and A. J. Wilkie. VI, 410 pages. 1980.

Vol. 835: H. Zieschang, E. Vogt and H.-D. Coldewey, Surfaces and Planar Discontinuous Groups. X, 334 pages. 1980.

Vol. 836: Differential Geometrical Methods in Mathematical Physics. Proceedings, 1979. Edited by P. L. García, A. Pérez-Rendón, and J. M. Souriau. XII, 538 pages. 1980.

Vol. 837: J. Meixner, F. W. Schäfke and G. Wolf, Mathieu Functions and Spheroidal Functions and their Mathematical Foundations Further Studies. VII, 126 pages. 1980.

Vol. 838: Global Differential Geometry and Global Analysis. Proceedings 1979. Edited by D. Ferus et al. XI, 299 pages. 1981.

Vol. 839: Cabal Seminar 77 – 79. Proceedings. Edited by A. S. Kechris, D. A. Martin and Y. N. Moschovakis. V, 274 pages. 1981.

Vol. 840: D. Henry, Geometric Theory of Semilinear Parabolic Equations. IV, 348 pages. 1981.

Vol. 841: A. Haraux, Nonlinear Evolution Equations- Global Behaviour of Solutions. XII, 313 pages. 1981.

Vol. 842: Séminaire Bourbaki vol. 1979/80. Exposés 543–560. IV, 317 pages. 1981.

Vol. 843: Functional Analysis, Holomorphy, and Approximation Theory. Proceedings. Edited by S. Machado. VI, 636 pages. 1981.

Vol. 844: Groupe de Brauer. Proceedings. Edited by M. Kervaire and M. Ojanguren. VII, 274 pages. 1981.

Vol. 845: A. Tannenbaum, Invariance and System Theory: Algebraic and Geometric Aspects. X, 161 pages. 1981.

Vol. 846: Ordinary and Partial Differential Equations, Proceedings. Edited by W. N. Everitt and B. D. Sleeman. XIV, 384 pages. 1981.

Vol. 847: U. Koschorke, Vector Fields and Other Vector Bundle Morphisms – A Singularity Approach. IV, 304 pages. 1981.

Vol. 848: Algebra, Carbondale 1980. Proceedings. Ed. by R. K. Amayo. VI, 298 pages. 1981.

Vol. 849: P. Major, Multiple Wiener-Itô Integrals. VII, 127 pages. 1981.

Vol. 850: Séminaire de Probabilités XV. 1979/80. Avec table générale des exposés de 1966/67 à 1978/79. Edited by J. Azéma and M. Yor. IV, 704 pages. 1981.

Vol. 851: Stochastic Integrals. Proceedings, 1980. Edited by D. Williams. IX, 540 pages. 1981.

Vol. 852: L. Schwartz, Geometry and Probability in Banach Spaces. X, 101 pages. 1981.

Vol. 853: N. Boboc, G. Bucur, A. Cornea, Order and Convexity in Potential Theory: H-Cones. IV, 286 pages. 1981.

Vol. 854: Algebraic K-Theory. Evanston 1980. Proceedings. Edited by E. M. Friedlander and M. R. Stein. V, 517 pages. 1981.

Vol. 855: Semigroups. Proceedings 1978. Edited by H. Jürgensen, M. Petrich and H. J. Weinert. V, 221 pages. 1981.

Vol. 856: R. Lascar. Propagation des Singularités des Solutions d'Equations Pseudo-Différentielles à Caractéristiques de Multiplicités Variables. VIII, 237 pages. 1981.

Vol. 857: M. Miyanishi. Non-complete Algebraic Surfaces. XVIII, 244 pages. 1981.

Vol. 858: E. A. Coddington, H. S. V. de Snoo: Regular Boundary Value Problems Associated with Pairs of Ordinary Differential Expressions. V, 225 pages. 1981.

Vol. 859: Logic Year 1979–80. Proceedings. Edited by M. Lerman, J. Schmerl and R. Soare. VIII, 326 pages. 1981.

Vol. 860: Probability in Banach Spaces III. Proceedings, 1980. Edited by A. Beck. VI, 329 pages. 1981.

Vol. 861: Analytical Methods in Probability Theory. Proceedings 1980. Edited by D. Dugué, E. Lukacs, V. K. Rohatgi. X, 183 pages. 1981.

Vol. 862: Algebraic Geometry. Proceedings 1980. Edited by A. Libgober and P. Wagreich. V, 281 pages. 1981.

Vol. 863: Processus Aléatoires à Deux Indices. Proceedings, 1980. Edited by H. Korezlioglu, G. Mazziotto and J. Szpirglas. V, 274 pages. 1981.

Vol. 864: Complex Analysis and Spectral Theory. Proceedings, 1979/80. Edited by V. P. Havin and N. K. Nikol'skii, VI, 480 pages. 1981.

Vol. 865: R. W. Bruggeman, Fourier Coefficients of Automorphic Forms. III, 201 pages. 1981.

Vol. 866: J.-M. Bismut, Mécanique Aléatoire. XVI, 563 pages. 1981.

Vol. 867: Séminaire d'Algèbre Paul Dubreil et Marie-Paule Malliavin. Proceedings, 1980. Edited by M.-P. Malliavin. V, 476 pages. 1981.

Vol. 868: Surfaces Algébriques. Proceedings 1976–78. Edited by J. Giraud, L. Illusie et M. Raynaud. V, 314 pages. 1981.

Vol. 869: A. V. Zelevinsky, Representations of Finite Classical Groups. IV, 184 pages. 1981.

Vol. 870: Shape Theory and Geometric Topology. Proceedings, 1981. Edited by S. Mardešić and J. Segal. V, 265 pages. 1981.

Vol. 871: Continuous Lattices. Proceedings, 1979. Edited by B. Banaschewski and R.-E. Hoffmann. X, 413 pages. 1981.

Vol. 872: Set Theory and Model Theory. Proceedings, 1979. Edited by R. B. Jensen and A. Prestel. V, 174 pages. 1981.